D0906667

NJBG
NEW JERSEY BOTANICAL GARDEN

NEW JERSEY STATE BOTANICAL GARDEN

SKYLANDS

The Garden of the Garden State

NJBG/Skylands Association
Ringwood, New Jersey

www.njbg.org

First published in the United States of America in 2011
by Adastra West, Inc.
Mahwah, New Jersey 07430

Skylands, The Garden of the Garden State / by the NJBG/Skylands Association, Inc.
ISBN 978-0-9674075-1-7

Cover: Swan Boy in the Perennial Garden
Title page: Grotto and pool in the Azalea Garden

Design by Maja Britton

Printed in the United States of America

Credits

Editors
Alex Rainer
Dorothy Dobek
Linda Glasgal

Writers
Nancy Bristow
Maja Britton
Richard Cahayla-Wynne
Ingeborg Langer
Dr. Edith Wallace

Photographers
Eugene Bigliano
Nancy Bristow
Maja Britton
Joseph Cooper
Daiv Freeman
Melina Fuda
Thomas Grissom
Gail Hawthorne
Edwin Kaar
Ingeborg Langer
Pauline Maniscalki
Peter Platt
Alex Rainer
Sharon Rounds
Dr. Edith Wallace
Jerome Wyckoff

NJBG Book Committee
Thomas E. Grissom, *Chair*
Maja Britton
Dorothy Dobek
Frank Dyer
Richard Cahayla-Wynne
Andrew F. Noll, III

Financial Support
The Fred J. Brotherton
Charitable Foundation
NJBG/Skylands Association

Ringwood State Park
Eric Pain
Superintendent
Richard Flynn
Head Landscape Designer

Contributors
Joanne Atlas
Senator Leanna Brown
Richard Cahayla-Wynne
Carnegie Library of Pittsburgh
Gail Doscher & Tom O'Brien
Richard C. Greene
Norma Herzfeld
Governor Thomas H. Kean
Library of Congress
Congressman Robert Roe
Timothy Roe
Borough of Ringwood
Linda Spirko
David & Lorraine Cheng Library
William Paterson University
of New Jersey
Robert Wolk
Jerome Wyckoff
Clare Yellin
Samuel Yellin Metalworkers

"When I designated Skylands as New Jersey's State Botanical Garden in 1984, it was my hope that this would help protect and preserve its unique collection of native and international plants, its magnificent Tudor landscape, and the natural beauty of the Ramapo Mountains.

Today that hope is fully realized. Thanks to the hard work of its dedicated State Parks staff, teamed with the selfless efforts of many hundreds of volunteers, Skylands is flourishing as a natural treasure and a priceless resource for all of New Jersey's citizens.

The thousands of visitors who find recreation and inspiration here each year prove that the designation was most appropriate. And their growing numbers give us confidence that the appreciation of this marvelous symbol of our State's rich natural and man-made heritage will continue to blossom in the years ahead.

Thomas H. Kean

Governor, State of New Jersey 1982–1990

Table of Contents

Preface

From its agrarian beginnings through the Gilded Age and into the modern era, Skylands has remained an oasis of beauty and serenity in the shadow of the Ramapo Mountains.

The story of Skylands' evolution from farmstead to National Historic Site contains equal parts labor of love and the foresight to preserve the gardens and buildings for future generations. It is a serendipitous blend of opportunity and destiny.

The NJBG Skylands Association saw the need to document this story. Through the generous support of the Fred J. Brotherton Charitable Foundation, this book became a reality to illustrate how Skylands became the Garden of the Garden State.

Thomas E. Grissom
NJBG Book Committee Chair

Above: The rolling Ramapo Mountains from the top of Mount Defiance. Skylands Manor is in the front center.

Right: Not far from the entrance to Parking Lot C you'll find a large leftover from the last Ice Age. This boulder was deposited near its current location about 13,000 years ago by a melting Pleistocene ice sheet. It is a rock fragment broken off by the advancing glacier from some hill to the northwest. It was rounded, polished and grooved by abrasion as the glacier moved it, along with other rock debris, to this place. Grooves are at an angle to dark streaks in the rock and are most noticeable when wet. The boulder is called an 'erratic' because it has traveled far from its place of origin.

The Ramapo Mountains

Skylands is located in one of North America's oldest landscapes—the Appalachian Mountains. This band of very ancient highland rocks extends from Maine to Georgia, and includes the Hudson Highlands and the Ramapo Mountains.

Once higher than the Alps, the Ramapos we see today are just the foundations of the imposing peaks they were long ago. Worn down by erosion during the remote geologic past, they were uplifted and reconfigured in relatively recent time. The latest major reshaping occurred during the last Ice Age, when the terrain was sculpted by the Laurentide Ice Sheet as it advanced to the south and east. The moving ice scoured the land, carving valleys deeper, plucking up boulders, and strewing them and the smaller glacial drift across the landscape.

After the ice retreated, these rugged hills became home to a sparse human population, first of woodland Lenape and much later, a scattering of colonial settlers. It was only after the American Revolution that the rolling Ramapos around Ringwood began their slow transformation into the treasure that is today's New Jersey Botanical Garden.

The region had played an important role in the nation's struggle for independence: the rich iron deposits of the Great Ringwood Tract had provided cannonballs for the Revolutionary cause, and Robert Erskine, lord of Ringwood Manor, served as Surveyor-in-Chief for General Washington.

But now that struggle was finally resolved, and it was time to beat the swords into plowshares. Now the land would yield agricultural bounty as well as mineral wealth.

In 1801 David Storm bought the first parcels of land for his Brookside Farm. Storm was soon joined by Joshua Morris, who acquired 33 acres of the Great Ringwood Tract in 1802. Both his Morris Farm and the Storms' Brookside Farm thrived and expanded in the course of the nineteenth century. Despite the rocky, uneven terrain, the land proved suitable for crops ranging from corn to hay, from beets to oats. By the time the two properties were sold to form the core of Skylands Farms, they comprised about 700 acres.

The sale of the Brookside Farm in 1891, and of half of the Morris Farm in 1892, marked the shift from family farming to operation as a "gentleman's

Left: As you stroll around the Botanical Garden, note the many ways in which native stone was used by the original planners of Skylands when they designed the property. It is the foundation of many of the Garden's Stetson-era buildings, such as the charming Pump House. There are many stone walls and terraces, as well as an entire drainage system of hand-laid stonework along the roads, in the Garden, and in the surrounding hills.

Left: Stone for the exterior walls of the Lewis Manor House was quarried at Pierson Ridge, above Emerald Pool, in the eastern part of the property that lies in Bergen County. The cutter was James McLaren & Sons, of Brooklyn. Architect John Russell Pope's specifications state that "all stones must be accurately cut [and] marked by section and course number, showing the exact place where each belongs. The Stone Setter will be held responsible if stones are taken from where they belong to be put in any other place."

farm." Such ventures into "scientific farming" had become a favorite pastime of wealthy financiers and industrialists toward the end of the 19th century.

In this case the "gentleman" was Francis Lynde Stetson, who took time out from his busy corporate-law practice to make the purchase. No matter how pressing his business affairs, Stetson managed regular returns to the Ramapo Mountains, to develop Skylands into the perfect summer retreat.

And develop it he did, commissioning a substantial residence and extensive farm buildings, and adding further parcels to his holdings. By the time of his death in 1920, he had expanded the estate to a total of more than 1,000 acres.

The Carriage House was built in the late 19th century with rocks hand-selected at local quarries. They represent a good cross-section of local geology. The front entrance is on the building's western face.

1. **Banded gneiss** (metamorphosed granite): Bands are edges of varied mineral layers
2. **Granite pegmatite** (igneous rock): Large crystals of feldspar (pink) and quartz
3. **Sandstone** (sedimentary rock): Formed from sand deposited in layers on streambed or lake bottom.
4. Mostly **hornblende** (common iron-bearing mineral)
5. **Granite**: mostly of quartz and hornblende
6. **Granite**: mostly of feldspar (pink) and quartz
7. **Granite**: mostly of hornblende (black) and feldspar
8. **Sandstone**: Grains strongly cemented by quartz deposited between sand grains
9. **Sandstone**: "Veins" are tunnels made by plant roots or burrowing animals in original sand, later filled with other sediments
10. **Conglomerate**: With white quartz pebbles. White streaks are tension cracks that filled with quartz from hot solutions at depth. Cracks formed when the rock was under tension during deformation. Quartz was deposited in the cracks by hot solutions at depth.
11. **Diabase** (iron-bearing igneous rock): Like Hudson Palisades rock. Formed from lava-like molten material breaking through or squeezing between layers of bedrock at depth.

Compiled by Jerome Wyckoff; © NJBG 2004

Francis Lynde Stetson

The Stetson Era

FRANCIS LYNDE STETSON

Francis Lynde Stetson (04/23/1846-12/05/1920) was born in Keeseville, in upstate New York near Lake Champlain. He was a descendant of the Massachusetts Lyndes and of Robert and Honour Stetson, who had come from England and settled in Scituate, Massachusetts, in 1634.

Francis was the third of four sons of Albany judge and U.S. Congressman Lemuel Stetson and his wife, Helen Hascall. Oldest brother Ralph died in 1859, when Francis was just thirteen; and second brother John, a Lieutenant Colonel with the 59th New York Veteran Volunteers, fell in 1862 at the Battle of Antietam. As the eldest surviving son, Francis was left with the responsibility of following in his father's distinguished footsteps, and he rose to the challenge manfully.

Stetson graduated from Williams College in 1867, and from Columbia Law School in 1869. For the rest of his life, he continued his affiliation with Williams College as Trustee and benefactor, as evidenced by the library building, scholarship and two campus roads named in his honor.

In 1873, he married Elizabeth Ruff of Rahway, NJ, but their marriage remained childless. In need of an heir after her death in 1917, Stetson adopted his wife's young secretary, Margery H. Lee, as his daughter.

Stetson was lead attorney of the law firm that evolved into the New York establishment of Davis, Polk, & Wardwell. From 1905 to1908 he helped write the country's first code of legal ethics, and he was President of the New York Bar Association in 1910-11. He was President Grover Cleveland's most intimate advisor, and Cleveland, between presidencies (1889-1893), was "of counsel" to Stetson's firm, then called Bangs, Stetson, Tracey, & MacVeagh.

Stetson's work with J.P. Morgan and Andrew Carnegie, during the "Golden Age" at the end of the 19th century, established him as a founder of modern corporate law. Indeed, his book, *Some Legal Phases of Corporate Financing, Reorganization and Regulation* (1922), can still be purchased today.

Stetson first met J. Pierpont Morgan in 1885, while working on a complex transaction involving the Pennsylvania Railroad. The two hit it off, and that relationship continued until Morgan's death in 1913.

J. P. Morgan

Working for and with Morgan interests, Stetson either directly or indirectly handled much of the most important railway litigation in the country. At various times he himself was a director of nine different railroads. At the time of his death he still served on the boards of the Erie Railroad; the New York, Susquehanna and Western; and the Chicago and Erie Railroad.

His interests extended well beyond railroads. In 1888 he initiated the construction of the first large hydroelectric power plant in America, at Niagara Falls, New York, and at the same time established the standard for the American electric system. Also in the late 1880s, he worked with William Whitney on the creation of the first integrated public transit system in New York City.

Stetson's legal expertise was instrumental in the formation of General Electric (1892) and of the International Paper Company (1898). And in 1901, as Morgan's lawyer, Stetson forged the legal links among the Carnegie Steel Corp., coal mines, iron ore mines and railroads that created the US Steel Corporation. With a capitalization of over $1.4 billion, it was the world's largest industrial combination up to that time, representing 7% of the country's gross domestic product.

Beyond all his business and legal accomplishments, Francis Stetson had a deep interest in the natural world as well. He was a member of the New York Botanical Garden's Board of Managers from 1908 until his death, and served as Vice President of the Garden beginning in 1914. He freely and generously contributed time and money to that institution's development. Indeed, the genus *Stetsonia*, a gigantic Argentine cactus, was named after him for his services to botanical science.

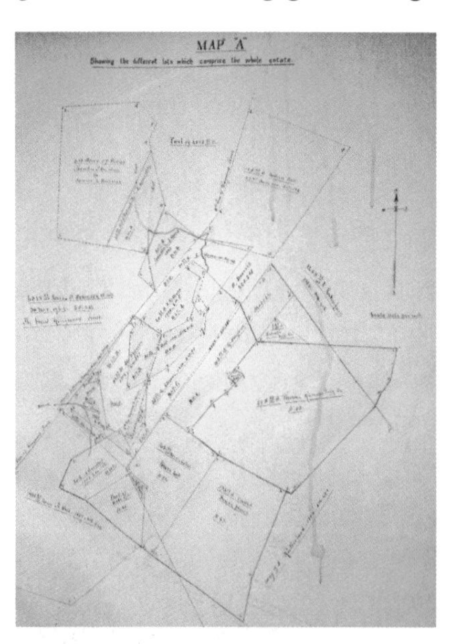

Stetson's master map of the parcels he acquired to create Skylands

But his interest in botany was not limited to the constraints of a formal botanical garden. As the Garden's Board of Managers stated upon his death, "Wild plants were of great interest to him and he was an enthusiastic advocate of the preservation of natural features and the conservation of natural resources."

At Skylands Farms, he was able to give free rein to this passion. On that estate, Stetson lavished the powers of his great mind and fortune. Skylands gave Stetson much pleasure, and he took delight in all of its aspects.

The lawn fronting Mr. Stetson's large, impressive mansion incorporated a portion of his nine-hole golf course, and here we see part of his herd of sheep busily keeping it well manicured.

FRANCIS LYNDE STETSON'S SKYLANDS FARMS

In 1891 Francis Lynde Stetson started acquiring farmsteads in the Ramapo Mountains, to transform them into the perfect summer retreat. But it took years to fully implement his vision and create the assemblage that his friend Andrew Carnegie called "the most beautiful country estate in America."

Stetson's Skylands included 250 acres of good farmland that produced corn, hay, timothy, beets and oats from soil described as light porous loam. The food crops yielded an ample supply for Mr. Stetson's family and guests and the resident employees. No fewer than 20,000 trees were set out during the years from 1909 to 1916. Many of the trees, especially five English yews, were rare and valuable. There were many sugar maples on the property and utensils for making maple sugar.

During the time of Mr. Stetson's ownership, the property was served by over twenty-eight miles of roadway. The drives were crown-surfaced with rolled crushed stone or gravel; they had drains, and gutters on both sides. There was maintenance equipment for these roads, and a source of stone and gravel nearby. There were also five miles of bridle paths, and these roads and paths were designed to serve as a firebreak from neighboring woodland. Wire fenc-

Oak Allee is one of the more than thirty miles of roads in Skylands. It was lined with trees, with a wire fence covered in honeysuckle. The fence was needed to enclose the pasture where sheep and cattle grazed. One of Parsons's principles was that if you must have a fence you should cover it with vines. Note the farm buildings at right.

The Oak Allee today with its mature trees forming a sunlit canopy over the roadway. The farm buildings and sheep enclosures have given way to meadows.

ing enclosed the estate, and various fields, paddocks and pastures were also fenced in. Fence lines were planted with honeysuckle or grapevine.

Skylands had an extensive water supply system: with a series of spring-fed ponds beginning at an elevation of 250 feet, it used gravity to bring water to all levels. The supply was sufficient for a small city and included hydrants for all buildings and other strategic points. To insure the safety of the water, there was a filtering system. Additionally, the Manor House and Lodge had dependable water from an artesian well and large pumps to bring drinking water to both structures.

At the center of Skylands Farms, Mr. Stetson had a mansion of native stone constructed. It was entirely surrounded by a nine-hole golf course, where sheep grazed when no golf game was in progress.

That gaslight-era mansion no longer stands, replaced in its exact location by the current manor house. But most of the outbuildings and many of the roadways throughout the property remain as they were in Stetson's day.

Stetson's manor house was completed in 1897. This adaptation of the Tudor style, built of fieldstone and covered in ivy, was the creation of architect Algernon E. Bell, who practiced in New York City from 1887 through the late 1910s. It included twenty-five rooms and nine baths, and was surrounded by velvety lawns. The lower floors of the house were sprinklered, and there was a complete fire alarm system. Like other buildings on the property, the manor house was gas-lit by a very safe and efficient arrangement, with gas delivered from a central lighting plant.

Close by the mansion, the Elizabethan-style Lodge, containing fourteen rooms, two baths, and a lavatory, provided additional housing for guests. Beautifully finished hardwood was chosen to complete its interior.

The layout of estate buildings was carefully thought out. Workmen were comfortably, sometimes even luxuriously, housed, and elaborate buildings were provided for all kinds of farm animals. There were various garages, sheds and shops, which made the estate self-sustaining.

For the design of a model farm contained within the confines of the Skylands property, Francis Stetson awarded the commission to S. Alfred Hopkins. He specialized in country houses and farms for the American elite, and he was known as the "dean of farm group architecture." Hopkins, who created farm groups for such clients as Louis Comfort Tiffany, sometimes designed ancillary buildings on an estate where the residence was the work of another prominent architect, in this case Algernon Bell.

Gatun, named after the great lake created during the building of the Panama Canal, was one of the two reservoirs in the higher elevations of the estate. It supplied water to the house and gardens by gravity. This photo, taken from the bathhouse put in by the next owner of Skylands, shows the pastoral view even this far from the residence.

The Lodge, behind an extensive lawn surrounded by young trees, is shown without the sundial designed by Clarence Lewis to enhance the large chimney. The two 'eyebrow' windows were also replaced with dormers when Lewis adapted the building.

Hopkins's farm buildings were practical, picturesque and esthetic, blending with the landscape and separating farming functions while all harmonizing with the manor house. Among the buildings at Skylands credited to Hopkins are the cow barns, dairy, the Lodge, coachman's cottage, stables, the home of the estate manager, the carriage house (today's Visitor Center), and the picturesque pump house.

The estate included living quarters for nine families in comfortable, modern buildings of stone, frame and stucco. The coachman's cottage, a beautiful stone building, contained fifteen rooms, and there were five concrete cottages to house seasonal laborers.

The equipment of the cow barns was the best that could be obtained; it included a heating system to be used in especially cold weather, and a modern ventilating system to continually supply pure air. The milk processing equipment was state-of-the-art, and dairymen enjoyed fine living conditions and sanitary provisions.

The poultry farm compared with the best commercial poultry farms in the country. This farm at Skylands could accommodate five hundred laying hens and included an incubator cellar that could hatch 3600 eggs at a time. A duck house on the shore of one of the ponds sheltered three hundred birds. There was also a piggery and a separate department for the sheep. The shepherd

Made of rough fieldstone and hewn timbers, the Coach Stable and Coachman's Cottage built for Francis Stetson still stands in the Botanical Garden. Today it is used as a residence for State staff.

The chicken houses, yards and brooder house were designed by Alfred Hopkins, among the most notable farm designers of the age. At right is a storage shed.

was provided with a heated six-room house that included a bath, plus hot and cold running water. The horse barn had room for eleven horses, and two box stalls as well. There were also buildings for food storage, and a smokehouse for processing hams and bacon. The icehouse had a capacity of eight hundred tons. Remarkably, even a saw mill could be found among the many estate structures. The builders of Mr. Stetson's Skylands Farms seem to have forgotten nothing.

The sheepfold in the early 1900s.

The Pump House in the snow. During Stetson's time, it served as an informal gatehouse for the property. In the left background you can see the greenhouse, a caretaker's house (now called the Bussink House), and a windmill.

The west view of Stetson's house photographed in 1916

Francis Stetson's guests at Skylands included (left to right) President Grover Cleveland, Ethel Barrymore, and Andrew Carnegie.

It is not difficult to imagine the social life enjoyed at Skylands by a man President Cleveland's biographer described as "one of the most cultivated lawyers in the city and a man of exquisite courtesy and social grace." Friends from the Century, Metropolitan, and other of Stetson's clubs often visited. Old-timers remember Ethel Barrymore and Andrew Carnegie. And of course Grover Cleveland and J.P. Morgan were frequent visitors.

They found it easy to reach Skylands, whose address at that time was "Sterlington, New York." Thanks to a road over the mountain, from the estate to the Erie station in Sterlington, one could leave the manor house in the morning and be at a Wall Street desk just a few hours later.

But despite that convenience, Mr. Stetson didn't travel to the City every day. With a good part of his mansion's ground floor dedicated to a comfortable study and generously sized library, we can assume that he spent some of his time at Skylands working on legal matters.

He had initially risen to prominence as a trial lawyer, but from 1898 on he focused almost completely on corporate work, which did not require court appearances, or rigid office hours. As his obituary stated, "He was a hard worker and did not go downtown or come home by his timepiece. There were days when he was not seen at his office at all, but those who knew him came to learn that when he was thus absent he was working elsewhere, and when he did return to the office it meant more work for everybody." ["Francis L. Stetson, Lawyer, Dies at 74," *New York Times* 12-6-1920]

Nonetheless, Mr. Stetson's life at Skylands was anything but drudgery. At the estate, he and his guests could enjoy fresh food and healthy outdoor activities. The fine nine-hole golf course was right outside the manor house. For sportsmen, there were deer and foxes in the surrounding woods, and the ponds and brooks teemed with game fish.

Skylands furnished an abundance of fruits and vegetables both in and out of season. The greenhouse plant was kept in perfect repair, and in that setting, fruits, flowers and vegetables were grown under glass.

Within view of the manor house was Mr. Stetson's formal garden, arranged for a continuous succession of bloom, and his rose garden contained a profusion of rare and precious bushes. A three-hundred-foot rose hedge, flanked by white peonies, enclosed the small fruit garden. The formal garden was furnished with a sundial and garden seats, and its gateway was marked by a pair of seventeenth-century Venetian stone dogs.

The design and layout of the grounds of the estate was the responsibility of Samuel Parsons, Jr. (1844-1923), a protégé (and later, partner) of Calvert Vaux, who together with Frederick Law Olmsted had designed New York City's Central Park.

Parsons, the son of Samuel Bowne Parsons and Susan Howland Parsons, was born in New Bedford, Massachusetts. His ancestors had emigrated from Somersetshire, England, to Pennsylvania. His father is remembered as the first American horticulturist to propagate rhododendrons and to import Japanese maples. The family's nursery in Flushing, Queens, which had gained international recognition, prospered until the elder Parsons's death in 1907.

After young Samuel graduated from Yale Scientific School, he began his career as a landscape architect. Having already gained practical experience and

Sterlington Station on Route 17 in Sloatsburg, New York.

Samuel Parsons

Samuel Parsons and a sample of his landscape design at Skylands

familiarity with plants and landscape materials at the family nursery, he lucked into the opportunity of a lifetime: Calvert Vaux offered to make Parsons his apprentice, and this relationship blossomed into a full-fledged partnership over time. When Vaux was named head landscape architect of the New York City Parks Department, Parsons came along as Superintendent of Planting. And after Vaux's tragic death in 1895, Parsons succeeded him as the City's head landscape architect, a position he held until 1911.

A founder of the American Society of Landscape Architects, Parsons is credited with the design of residential projects in at least thirteen states, as well as parks throughout the country. He wrote extensively during his career, and his philosophy on design and his distinctive influence on the American landscape can be learned from his numerous publications. In the most widely read of these, *The Art of Landscape Architecture* (1915), it is evident that Skylands held a special place in Parsons's heart. Based on a gentleman's working farm of the period, Skylands features prominently in the photos included in that book to illustrate his design ideas. And though others expanded on Parsons's design in later years, we can still see his imprint in the Skylands of today.

After completion of his mansion in 1897, Francis Stetson had twenty wonderful years in which to enjoy his country estate. It was a perfect place to be soothed by the tranquility of nature, to pursue his legal responsibilities in a calm environment far from city stress, and to meet with clients and friends in a relaxed, congenial setting. But it could not last forever.

Left: The children of estate manager, James Kendall, enjoying a ride in a wagon just the right size for them. This photo was taken circa 1916-1920.
Right: Clarissa Wardwell Pell, daughter of one of Stetson's business partners, was a frequent visitor at Skylands. She is shown here at age 6 on the circular front drive, circa 1915-1916.

In April of 1917 came the first blow, with the passing of Elizabeth Ruff Stetson, his wife of 44 years. Then, in 1918, Mr. Stetson himself suffered a stroke. He remained in ill health thereafter, often confined to his New York City home on West 74th Street. Now it was impossible to recapture the gaiety and conviviality of his earlier years in the Ramapo Mountains.

If such good times were to return to Skylands, it would have to be with a new lord of the manor. That transition was initiated by Francis Lynde Stetson's death in December 1920, which opened the way to new ownership—and a new manor house.

Some of Stetson's farm equipment is still on display today at the end of Oak Allee.

Clarence McKenzie Lewis

The Lewis Estate

The next master of Skylands was equally as enthusiastic about horticulture as Francis Stetson had been—or perhaps even more so. He maintained some of Stetson's estate virtually intact, while making a dramatic change at the very heart of the property.

The new owner was Clarence McKenzie Lewis—civil engineer, railroad-man, banker and passionate plantsman—who purchased Skylands in 1922.

Buying the estate was a natural choice. Mr. Lewis was already familiar with the Skylands property and all that it contained, especially the gardens.

One reason for his familiarity lies in livestock. On his Sheffield-Hope Farm in nearby Mahwah, Lewis raised Suffolk Punches, a breed of heavy English draft horses also favored by Francis Stetson. Visits bringing Sheffield Farm mares to be serviced by Skylands stallions provided the perfect opportunity for Clarence Lewis to get to know Skylands better—and made this a logical purchase when he decided to replace the Mahwah property.

But just who was this new owner of Skylands?

Born October 26, 1876, in Jersey City, Clarence McKenzie Lewis was a lonely only child who lost his father early. When he was nine years old, he was sent abroad to be educated in Germany and England. In 1892, while he was still in Europe, his widowed mother, Helen Forbes Lewis, married William Salomon, founder of a New York banking house and direct descendant of Haym Salomon, who had helped finance the American Revolution.

Clarence McKenzie Lewis and his mother, Helen Forbes Salomon

Upon returning from Europe, Lewis lived with his mother and stepfather at 1020 Fifth Avenue and attended Columbia University, where he received a Civil Engineering degree in 1898. He then moved to Cincinnati and went to work as an assistant chief engineer for the Baltimore and Ohio Railroad, riding inspection cars and checking on routing and bridges. Here he invented a code for the B&O which he later developed into an international banking code used extensively during the 1920s.

Clarence McKenzie Lewis adored his wife Annah Ripley Lewis; this photograph was probably taken at their home at Sheffield-Hope Farm in Mahwah (circa 1915).

When he left the B&O and joined William Salomon & Co., he became the in-house expert on engineering and railroad projects for the banking house.

In 1908 Lewis married Annah Churchill Ripley, who was born in 1887 in Montclair. In 1912 they bought a large country place in Mahwah they called Sheffield-Hope Farm. Despite her rheumatic heart, Mrs. Lewis enjoyed gardening there, developing a rose garden and a dell garden. It was here that horticulture captured Lewis's fancy.

Sadly, Annah Lewis died in 1918, leaving Clarence to raise their two young children—"Mac," age 7 and Barbara, age 4—alone. It is said that he never recovered from her loss.

The next family tragedy came in 1919, with the death of Lewis's stepfather, William Salomon. His house at 1020 Fifth Avenue was to be sold and then razed. Many of the items in that house would be auctioned, but there were some that Mrs. Salomon wished to preserve at another site.

Together, Mr. Lewis and his mother decided to purchase Skylands from the Stetson estate and make it a showcase for their interests, horticulture and antiquities, as found in the manor houses of Europe.

Lewis commissioned Heinegke and Smith to design a series of small glass vignettes. One featured him, another showed his children, Barbara and Mac, and yet another portrayed their governess in Tudor dress.

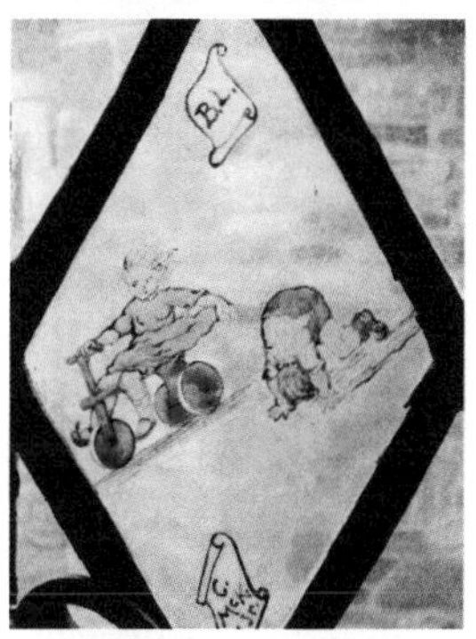

Gregarious, highly cultivated and extremely knowledgeable about furniture and the decorative arts, Mrs. Salomon wanted a Tudor showplace; her son wanted plants and gardens. The result was Skylands.

They embarked on this project by hiring John Russell Pope (1874-1937), one of the most prominent architects of that time, to design a fine Tudor Manor House, basically of the late Gothic/Renaissance period, to be situated at the same location as Mr. Stetson's home. (The basement of that earlier house remains, but it has been expanded to accommodate the current manor, which is considerably larger than the Stetson mansion had been.)

Mr. Pope was already familiar with the area: his first professional commission, in 1895, had been to design the chapel at Shepherd Lake in memory of Alfred L. Loomis, J. Pierpont Morgan's personal physician.

For the new Skylands Manor, Pope worked with the builder, Elliot C. Brown of New York City. He also brought in the firm Heinegke and Smith of New York City to produce the leaded windows which were to contain a collection of stained-glass medallions (primarily 16th century). Most of these medallions originated in Germany, Bavaria and Switzerland, although Mrs. Salomon made the purchase from an English collector. For inclusion in the Manor's windows Mr. Heinegke also designed thirteen unique glass pieces that depict family, workers, animals and plantings found on the estate.

James McLaren & Sons of Brooklyn, New York, was chosen to handle cutting and marking the granite for the house, including careful designation of the intended location for each slab. The stone was cut on the Skylands property at Pierson Ridge, near Emerald Pool.

Samuel Yellin of Philadelphia and New York was contracted to produce the entire wrought-iron exterior lighting, and he also designed much interior wrought-iron work as well, including the railing on the upper portions of the circular staircase.

Meanwhile, to design the gardens, Lewis had engaged the most prominent landscape architects of his day, the firm of Vitale and Geiffert.

The Manor House was on a knoll "looking as if it might slide off," so one of the first tasks was to build the West Terrace. Regiments of Model-T trucks brought soil up from Ringwood and Sloatsburg.

For the actual gardens, Lewis worked with Geiffert, and his own input was considerable. Lewis suggested the Crabapple Vista to transform the barren golf course. In the Octagonal Garden Lewis placed his rock plants, dwarfs and alpine flowers. Instead of the lindens Geiffert wanted in the Magnolia Walk, Lewis opted for a southern species, sweet bay magnolia, trained as trees, unusual for their size and number this far north. Lewis chose the locations for the Bog Garden and Wildflower Garden. For the latter Lewis wanted a "miniature New Jersey Pine Barrens," having soil brought in from the Barrens and then putting in appropriate plants. Lewis also originated the Lilac Collection and the Winter Garden, which holds New Jersey's largest Jeffrey Pine. Skylands contains four other official "grand trees" of New Jersey.

Vistas are a major feature of the Skylands landscape. Lewis had a great feeling for space. Every window has a view. Every vista has a terminus. A sculp-

Lewis enjoyed walking his property and sharing it with his guests.

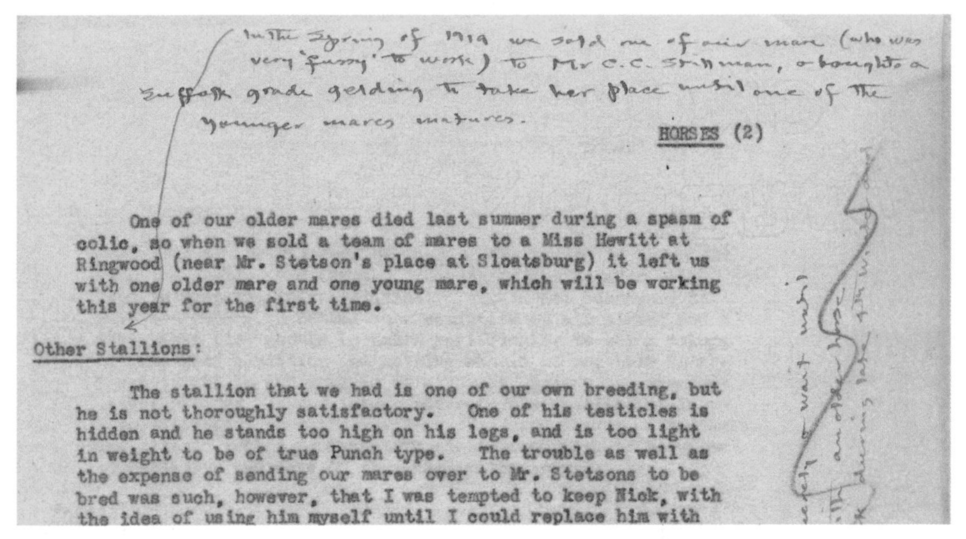

In the Spring of 1919 we sold one of our mares (who was very 'fussy' to work) to Mr C. C. Stillman, & bought a Suffolk grade gelding to take her place until one of the younger mares matures.

HORSES (2)

One of our older mares died last summer during a spasm of colic, so when we sold a team of mares to a Miss Hewitt at Ringwood (near Mr. Stetson's place at Sloatsburg) it left us with one older mare and one young mare, which will be working this year for the first time.

Other Stallions:

The stallion that we had is one of our own breeding, but he is not thoroughly satisfactory. One of his testicles is hidden and he stands too high on his legs, and is too light in weight to be of true Punch type. The trouble as well as the expense of sending our mares over to Mr. Stetsons to be bred was such, however, that I was tempted to keep Nick, with the idea of using him myself until I could replace him with

Lewis kept meticulous records of his activities at all of his properties, often correcting and annotating drafts in his tiny, precise handwriting. His logbooks still serve as a guide to Botanical Garden development.

ture, a flowering tree, a rock, a building draws the eye. In garden after garden, there are no dead ends in any season. This required ingenious planning and helps make Skylands unique.

Lewis had loved his railroad days, and as he worked with his gardeners he insisted on gentle curves, no sharp angles or curves in his flower beds. A frequent exclamation gardeners heard from him was, “Do you think a train could get around that? A train would never make it!”

His attention was caught by what was unusual, or what would be in “the wrong climate” for Skylands. After that came his concern for the final, overall effect.

New Jersey’s foremost horticulturist in his day, Lewis kept meticulous logbooks of every plant. Along with the hundreds of pages of technical data, origins and bibliographies, the logbooks contain detailed references for individual plants. These range from “seed collected from a tree about 35 to 40 feet high, and over 16 inches in diameter on the mountainside near the Chieh Lai Temple in China,” to “a fern with an entirely different odor found by Mrs. W. K. DuPont in the garden of an old house.”

Lewis collected specimens throughout the U.S., England and Europe, and beyond. Nearly every summer he went to the High Alps in Switzerland with his young children, who looked on while he clambered around inspecting his favorite alpine plants. He went to Alaska to see if these same plants grew at the same altitudes there. But one of his favorite places was the New Jersey Pine Barrens, especially around Whitesbog. He also found plants in other

From this 1928 photo of Magnolia Walk under construction, it is hard to imagine the stately effect created by the mature trees today.

New Jersey locations, from Bennett Bogs to Barnegat to Island Beach and Cape May. Other collecting trips took him farther afield, to Long Island, Cape Cod and beyond.

In 1929 Lewis accepted membership in the corporation of the New York Botanical Garden. Like Francis Stetson before him, he became a member of the Board of Managers. He was still a Trustee when he died.

Skylands was registered as a plant nursery and inspected twice a year by the U.S. Department of Agriculture. Lewis imported plants from such remote places as Afghanistan, the Azores, the Belgian Congo, New Zealand, Chile, and Kashmir. He was well known to nurserymen, collectors, arboretums and botanical gardens throughout the world, as well as to local plantsmen, zealously exchanging plants and information. His "idea of Heaven," it is said, was to join in the talk of the Head Gardeners from Planting Fields, Old Westbury, Wave Hill and other major gardens when they came together occasionally to talk shop at Skylands.

Skylands was a hospitable meeting place for horticultural societies and enthusiasts as well. Even as Lewis was preparing to leave Skylands, Thomas H. Everett, then director of the New York Botanical Garden, headed a last visit to Skylands and a testimonial dinner "for a group of appreciative garden enthusiasts . . . cognizant of the great contribution that Mr. Lewis has made to horticulture," as Everett put it in his invitation.

Other Skylands visitors in the Lewis era included such diverse persons as Jean Monnet, the father of the European Common Market; Sir Joseph Duveen, the art dealer; and Mme. Frances Alda of the Metropolitan Opera. Raymond Ditmars, renowned herpetologist with the American Museum of Natural History, came often and brought his "snake bag" on hikes. Lewis also welcomed fellow railroadmen: William Poland, who engineered the Trans-Alaska Highway and the railroad from the Persian Gulf to Teheran; and George Kittredge, former chief engineer of the New York Central Railroad, who shared Lewis's interest in racing pigeons.

But then came the day in 1953, when, at the age of 77, he no longer had the strength or resources to operate Skylands, and he sold the property to the National Bible Institute. This organization thought the property the ideal locale for the campus of its Shelton College.

After the sale was consummated, Lewis came back periodically to see the gardens. But Shelton let them deteriorate, keeping only one gardener where Lewis had employed sixty or more.

Eventually Mr. Lewis could not bear to see the property any longer. He died in 1959 in New York City, and is buried near his beloved gardens at Skylands, in the family vault in Saint Elizabeth's Memorial Chapel churchyard in Tuxedo, New York.

Clarence Lewis, his son and grandson on the West Terrace at Skylands Manor (top), and another photo of Lewis in younger days.

Above: The Crabapple Vista when it was first planted. Notice the Four Continents statues at the end.

Left: The Crabapple Vista today, as seen from the upper windows of the Lodge, its northern terminus.

Below: The Azalea Garden has an open, airy feeling in this early view. Over the years, the trees on either side have grown into a large and stately border.

The Landscape Architects

Ferruccio Vitale

In 1923, Clarence Lewis selected landscape architects Vitale and Geiffert to design his gardens. Such was the reputation of the firm that photos of their work were used to illustrate the *Encyclopedia Brittanica* article on Landscape Architecture.

Florence-born Ferruccio Vitale (1875-1933) first came to the United States in 1898, as military attaché at the Italian Embassy. After studying landscape design in Florence, Turin, and Paris, he came back to the United States in 1904. He soon joined forces with George Pentecost, Jr., who had been a partner of Samuel Parsons, Jr. and a co-founder of the American Society of Landscape Architects.

Elected to membership in that Society in 1908, Vitale was known primarily for his work in many private estates throughout the country, and on numerous public projects. Among the latter were Meridian Hill Park in Washington, DC, the town plans of Scarsdale and Pleasantville, New York, and the "Century of Progress" exposition in Chicago.

After Vitale's untimely death from pneumonia in 1933, his junior partner Alfred Geiffert, Jr. (1890-1957) continued his work. Geiffert, who worked closely with Clarence Lewis at Skylands, also created landscape designs for the National Gallery of Art, New York City's Rockefeller Center, and the 1939 New York World's Fair. He was Princeton University's landscape architect from 1943 to 1957.

The May, 1921, issue of *Bulletin of the Garden Club of America* singles out Geiffert's skill in placing "great natural boulders...in the most masterful manner." According to the *Bulletin*, "true Nature lovers owe [Geiffert] a vote of thanks for showing us how the wild woods can be brought to our very doors."

The Waterbearers in the Annual Garden

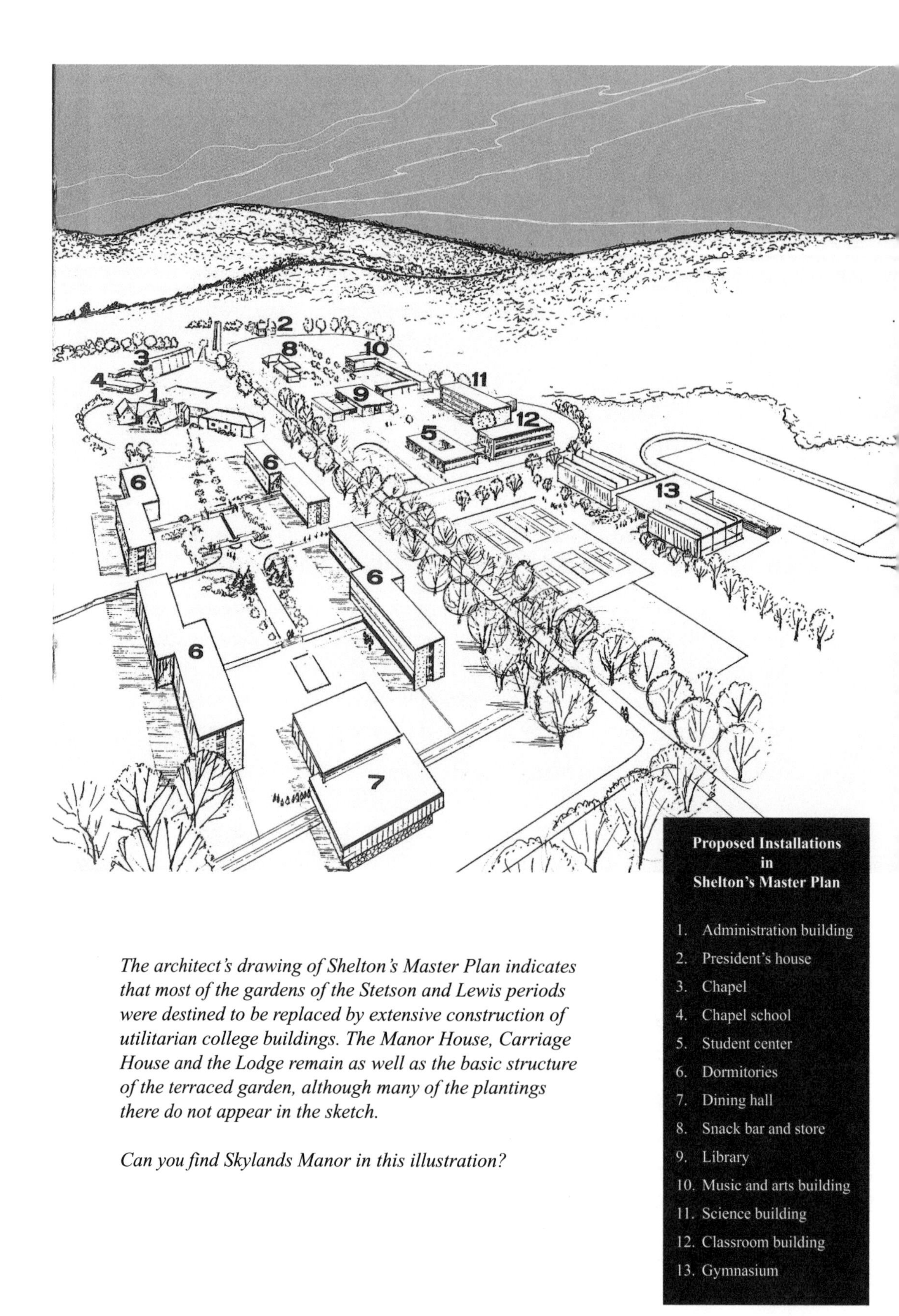

Proposed Installations in Shelton's Master Plan

1. Administration building
2. President's house
3. Chapel
4. Chapel school
5. Student center
6. Dormitories
7. Dining hall
8. Snack bar and store
9. Library
10. Music and arts building
11. Science building
12. Classroom building
13. Gymnasium

The architect's drawing of Shelton's Master Plan indicates that most of the gardens of the Stetson and Lewis periods were destined to be replaced by extensive construction of utilitarian college buildings. The Manor House, Carriage House and the Lodge remain as well as the basic structure of the terraced garden, although many of the plantings there do not appear in the sketch.

Can you find Skylands Manor in this illustration?

Shelton College

Shelton College National Bible Institute originated in New York City in 1908. In 1953, Dr. J. Oliver Buswell, college President from 1941 until 1956, moved the college to Ringwood, New Jersey, and Skylands became Shelton College - Skylands Campus. When Dr. Carl McIntire gained control of Shelton College, he transformed it into a Liberal Arts college. President McIntire resided in the Manor House.

The college added the annex now used as a dining facility by a caterer/bed and breakfast operator who has obtained a lease from the State of New Jersey for this purpose. At the time of the college's occupancy, the lower floor was utilized as a dining hall for students, and kitchen facilities were also located there. Daily chapel services and some classes were held on the upper floor of the annex. Occasionally, classes were held in the Great Hall of the Manor House, but that house was used primarily for administrative purposes. A former student reported that some seniors were given the privilege of housing on the third floor of the building.

Shelton College's brochure invited students to a unique learning environment.

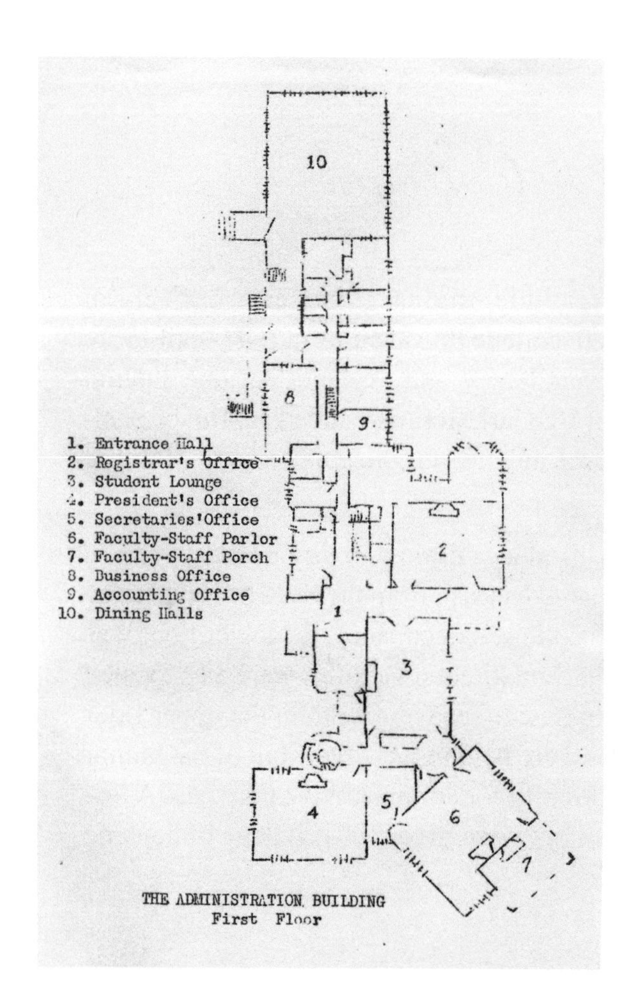

Left:
1. Entrance Administration Bldg. (Manor House), 2. Registrar's Office (Dining Room), 3. Student Lounge (Great Hall), 4. College President's Office (Library), 5. Secretaries' Office (Study), 6 & 7. Faculty/ Staff Lounge (Drawing Room and Porch), 8. Business Office (Staff Dining Room), 9. Accounting, 10. Dining Facility (Annex)

Below:
1. Administration Building, 2. Chapel/Music Studies (Carriage House), 3, 4 & 5. Classrooms/Women's Dormitories/Library (Hillair), 6. Women's Dormitory (The Lodge), 7 & 8. Proposed Gym and Athletic Field (east of Crabapple Vista), 9. Science Building (Dairy Barn), 13 & 14. Tennis Court/Greenhouses, 15, 16 & 17. Faculty/Student Housing (Hillair, East & Cascade Cottages), 18. Superintendent's Cottage (unchanged), 21, 22 & 23. Proposed Athletic Field/Dormitories

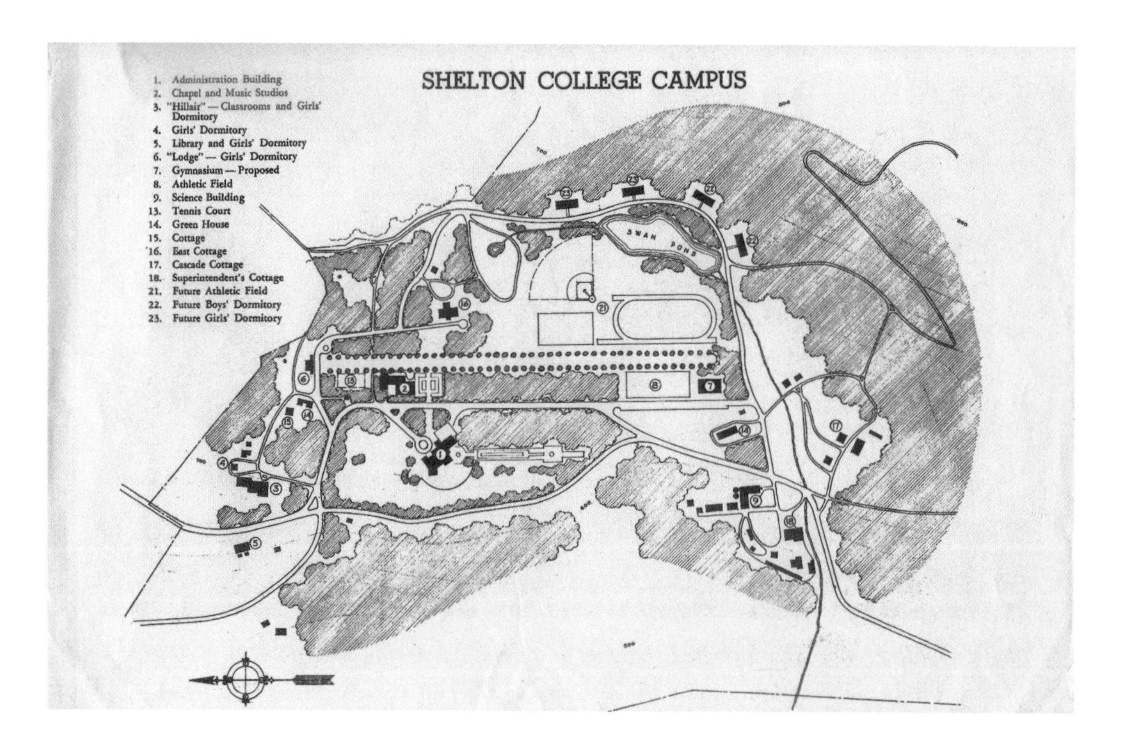

This student, who attended Shelton for one year (1959-60), also reported that a two-year Liberal Arts program was offered by Shelton and that most classes were held in the Hillair residence, the "big house" later destroyed by fire. During that time, there were no animals on the campus site. Sheep could be seen in a distant field, but they did not belong to the College.

Student housing for males was located in the old chicken coops, and female students were housed in various buildings on the Skylands property. Some of these structures are no longer standing. The building known as The Lodge was a dormitory for women. According to a former student, there were some small cottages on Culver Lake Circle available for occupancy by married students. These were leased for the "nominal" fee of $48 per month. A board member of the NJBG/Skylands Association related that his parents' summer cottage was leased for faculty housing during the school term.

In May of 1964, the college relocated to Cape May, New Jersey, and abandoned the Skylands Campus. The college continued to operate at the Cape May location until the 1980s, moving from there to a Florida campus. After that final move, the college closed, and it is no longer in existence.

Shelton College published its own student newspaper, as well as guides to both the campus and the surrounding gardens.

Shelton students were offered a broad range of liberal arts and religious classes, taught in the wonderful surroundings of the Manor House.

STATE OF NEW JERSEY
EXECUTIVE DEPARTMENT

EXECUTIVE ORDER NO. 62

WHEREAS, the Skylands Botanical Gardens, an area of approximately 80 acres located in the Skylands Section of Ringwood State Park, contain a unique collection of woody and herbaceous plants, flowers and shrubs; and

WHEREAS, the Skylands Botanical Gardens represent the finest collection of native and international plants, flowers and shrubs found in any State park or State-owned botanical area in New Jersey; and

WHEREAS, the Skylands Botanical Gardens are a major attraction to thousands of visitors to Ringwood State Park and hold a treasure of rare and exotic specimens for scores of plant collectors and botanists; and

WHEREAS, these gardens are integrated into a magnificent tudor landscape of fountains, stone masonry, and the natural beauty of New Jersey's Ramapo Mountains; and

WHEREAS, special emphasis placed upon these outstanding botanical gardens would serve to further enhance the image of New Jersey and would encourage the preservation, protection, documentation and display of the gardens as a symbol of our State's natural and man-made heritage;

NOW, THEREFORE, I, THOMAS H. KEAN, Governor of the State of New Jersey, by virtue of the authority vested in me by the Constitution and laws of the State of New Jersey, do hereby ORDER and DIRECT:

1. The Skylands Botanical Gardens are designated as the New Jersey State Botantical Gardens.

2. The care, protection and future of this rich heritage, a task of concern and importance to the people of this State, is continued as a responsibility of the New Jersey State Park Service.

3. This Order shall take effect immediately.

GIVEN, under my hand and seal this sixth day of March in the year of our Lord, one thousand nine hundred and eighty-four, of the Independence of the United States, the two hundred and eighth.

NEW JERSEY 1776 EXECUTIVE

Th H Kean
GOVERNOR

Attest:

Chief Counsel to the Governor

The Green Acres Purchase

During Richard J. Hughes's term as the Governor of New Jersey, Robert A. Roe held the position of Commissioner of the New Jersey State Department of Conservation and Economic Development. In 1966, Commissioner Roe announced the purchase under the Green Acres Program of the former Shelton College property located in Ringwood Borough, Passaic County, and Mahwah Township, Bergen County.

A November 23, 1966, copy of the *Home and Store News*, a weekly publication of the Journal Publishing Co. of Ramsey, New Jersey, reported the purchase price to be $2,217,488, and described the Skylands property as it was developed under the ownership of Clarence McKenzie Lewis.

The newspaper article further stated that "the preservation of this area is an integral part of the master plan for the Ringwood State Park expansion program." U.S. Secretary of the Interior Stuart L. Udall had recently declared Ringwood Manor a National Landmark.

Commissioner Roe stated in the credited publication, "I regard the preservation of this area of the state as essential and a foremost opportunity to meet the Green Acres objectives. The area is directly within the sweep of the population growth of our northeastern counties and communities. The Ringwood State Park conservation expansion program preserves our heritage and the natural beauty of our state."

Left to right: Gathered at the signing of the Executive Order creating the State Botanical Garden were Ringwood Mayor Walter Davidson, Ringwood Clerk Catherine Senicola, Parks & Forestry Director Greg Marshall, Governor Thomas H. Kean, Mrs. Hans Bussink, Senator Leanna Brown, and Mrs. J. Duncan Pitney.

New Jersey State Department of Conservation and Economic Development Commissioner Robert Roe (second from left) was an ardent supporter of the State's Green Acres program, and in 1966 spearheaded the purchase of the property that became the New Jersey State Botanical Garden. He is shown here with Governor Richard J. Hughes (second from right), during whose tenure the Shelton College property was acquired. Commissioner Roe later served as a New Jersey Congressman.

Left to right: At the Botanical Garden's dedication ceremony in 1985, guests included Passaic County Freeholder Richard DuHaime, Ringwood Mayor Walter Davidson, Governor Thomas H. Kean, and Senator Leanna Brown.

“ The designation of Skylands as the State Botanical Garden was the result of years of dedication by people who loved the Garden.

Governor Thomas H. Kean needed much persuasion to make the designation in 1984. He feared that eventually there would be demands for the state to contribute to Skylands and, like today, public funds were scarce. He could not resist the efforts, however, of so many citizen volunteers who advocated tirelessly for the designation.

In 1985 the Governor came to dedicate the Garden. The next year we had a press conference to broaden awareness of the great beauty of the property which had been 'saved from going to pot,' a phrase used often by my office assistant Isabelle Cunningham, who spent countless hours in the effort to save Skylands.

Thanks again to much hard work by garden-club members such as Mrs. J. Duncan Pitney of Mendham, countless volunteers and the staff at Skylands, and the New Jersey Park Service led by Gregory Marshall, a bill, which I sponsored, to appropriate $300,000 was introduced and passed in 1987.

The Garden and the entire 96 acres of land were preserved in 1965 by the foresight of Robert Roe, Commissioner of the New Jersey Department of Conservation and Economic Development and a Passaic County Freeholder. The area is truly a sparkling jewel in the crown of the Garden State.

Leanna Brown
NJ State Senator, 1984-93
NJ Assemblywoman, 1980-1983

Left to right: In 1985, Ringwood Mayor Walter Davidson, Senator Leanna Brown and Governor Thomas H. Kean displayed a newly-commissioned map of the Botanical Garden. The map is shown below.

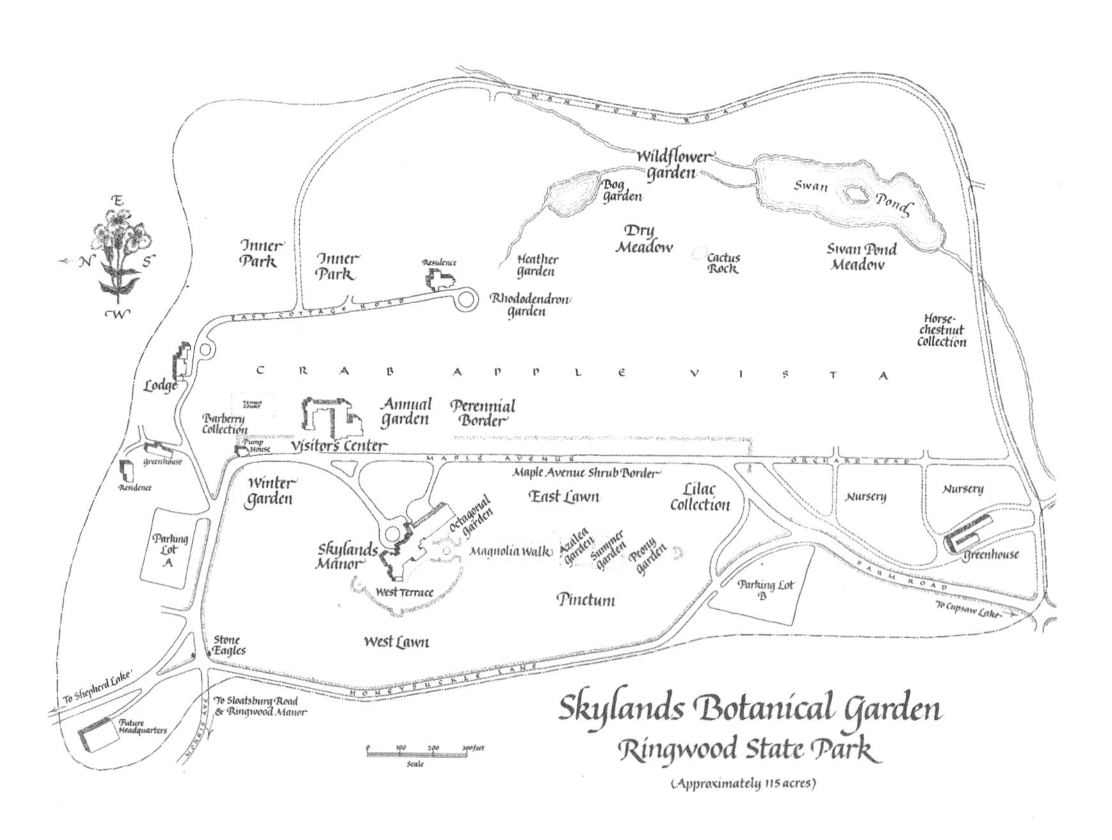

Left to right: Twenty-five years later, in 2010, guests at the 25th Anniversary Rededication included Mahwah Regional Chamber of Commerce President Annette Freund, Head Landscape Designer Rich Flynn, NJBG President Thomas Grissom, Matthew Weiss representing the office of Congressman Scott Garrett, Senator Leanna Brown, and Borough of Ringwood Manager Kelley Rohde. Ms. Rohde read a Proclamation from the state legislature honoring the Botanical Garden. The guests are gathered next to the official dedication stone.

Today, the State Park system maintains the Botanical Garden with a hardworking botanical staff under the direction of Head Landscape Designer Rich Flynn (third from right). This tireless group of state gardeners forms the backbone of the Garden's team, and today they take on the work done by dozens of gardeners when Skylands was a private estate. And it shows! Directing a small group of seasonal workers and hundreds of volunteers from the NJBG/Skylands Association, the regional horticultural community, and local businesses and civic groups, they keep the Botanical Garden blooming beautifully all year round.

Each year NJBG honors volunteers whose service goes above and beyond the call of duty. Many contribute hundreds of hours to keep the Botanical Garden blooming and to help the Skylands Association serve the Garden's thousands of visitors.

The Carriage House Visitor Center is operated by NJBG volunteers and staff.

The Skylands Association

In 2011, the NJBG/Skylands Association celebrated its 35th anniversary serving the Botanical Garden and the people of New Jersey.

The State bought Clarence McKenzie Lewis's Skylands estate in 1966 as its first purchase under the Green Acres law, which preserves land as public open space, and Skylands became part of the State park system. With no admission charge to offset the enormous cost of running such a large garden, local volunteers pitched in to help maintain the property.

In 1976, these hardy volunteers banded together to form the NJBG/Skylands Association, and became the Garden's official state-recognized 'Friends of the Park' organization. As such, the group expanded its mandate beyond gardening and began providing services for visitors. By the time Governor Thomas Kean dedicated Skylands as New Jersey's official botanical garden in 1985, the Association was in full bloom.

Today, NJBG continues to work with the State to transform this beautiful nineteenth-century estate into a world-class botanical garden. Each year the

Celebrating the groundbreaking at the Botanical Garden's Carriage House Visitor Center, were (left to right) Head Landscape Designer Richard Flynn, NJBG VPs John Gall and Thomas Grissom, Department of Parks and Forestry Director Jose Fernandez, Ringwood State Park Superintendent Rebecca Fitzgerald, NJBG president Diana Merkel, and Carriage House restoration architect Jon Fellgraff.

The spring Plant Sale gets the season off to a colorful start. It is a favorite among garden aficionados throughout the region.

The Lilac Garden gets special care from teams of volunteers under the supervision of a renowned regional expert. Classes like these combine garden maintenance and improvement with hands-on learning opportunities for Garden visitors.

Association fields hundreds of volunteers who work indoors and out, summer and winter. In 2010 alone they provided over 10,000 hours of effort, which the State valued at more than a quarter million dollars. Teams of NJBG volunteers help maintain the individual specialty gardens, have created several new ones, and have worked to conserve and improve prize plant collections such as the Lilac, Perennial, Wildflower and Hosta Gardens.

To educate and entertain the Botanical Garden's many visitors, NJBG maintains a full schedule of tours, walks, workshops, lectures, and large special events throughout the year, all managed and executed by NJBG volunteers. The spring Plant Sale, the summer Concert Series, the Harvest Fest, and the glittering Holiday Open House, each run by NJBG, draw thousands of new guests into the garden each year. There's a full operational staff of volunteers as well, people working quietly behind the scenes to run the gift shop, the botanical library, the website, marketing, publications, financial affairs and office management.

NJBG's work is underwritten by membership dues, donations, grants and contributions from civic-minded patrons and businesses. The Association receives no funding from the State or from on-site operators, and relies on the generosity of friends to support its ongoing projects.

NJBG's greenhouse is wall-to-wall with lush, healthy plants grown by volunteers for sale at the Plant Sale and for use in the Botanical Garden.

The summer concert series is a huge crowd-pleaser, with performances on the Concert Lawn all summer long. These Friday night delights feature a wide range of musical styles and appeal to all ages, with children taking advantage of the wide open play space while their parents enjoy the musical offerings.

Young and old alike turn out for Garden Cleanup days, where cooperation and companionship make the work go quickly.

When Clarence Lewis created the gardens at Skylands, he employed a staff of 60 full-time gardeners. Today, State gardening staff numbers in the single digits, and the Botanical Garden relies more than ever on the excellent Friends organization that NJBG has become.

How do we thank the countless hardworking and dedicated people who have given so generously of themselves for the last thirty-five years? Well, that's easy: we do it by building on their efforts to create an even better garden in years to come!

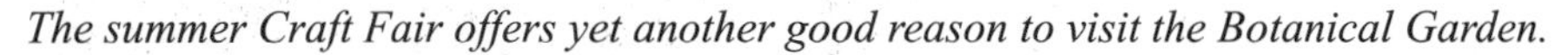

The summer Craft Fair offers yet another good reason to visit the Botanical Garden.

Hayrides in the woods and lots of colorful paint for decorating pumpkins are sure signs that the Harvest Fest has arrived. With good food, music, a craft fair, and lots of games for the youngsters, this event grows with every successive year.

The annual Holiday Open House, an NJBG tradition, is the grand finale of the Association's event-packed year. With the help of regional garden groups and clubs, as well as the generosity of area businesses, Skylands Manor is decorated floor to ceiling to welcome thousands of visitors.

As the sun sets, the Manor sparkles and glitters with lights in preparation for the festive evening Champagne and Candlelight tours.

Taking a Walk

NJBG offers walks, hikes and free garden tours throughout the year, with something for everyone – in every season.

Opposite page: A winter birding hike (top) and the early spring tree walk

This page, top to bottom: guided garden tour, Manor tour, family woodland walk

Skylands Manor

Clarence Lewis's purchase of Skylands was consummated on April 16, 1922, when his deed to the property was filed. For the next five years, the estate was a beehive of activity.

Lewis and his mother, Helen Salomon, had already concluded that the extant residence would not serve their purposes. Thus the first order of the day was demolition of Stetson's stone structure, to make way for the manor they had in mind.

This was to be not merely a grand summer home, but also a custom-made repository for Mrs. Salomon's extensive collection of antique interiors and *objets d'art*. Some of these items came from the Fifth Avenue mansion she had shared with her late husband William Salomon; others she acquired on collecting trips while the new Skylands Manor was being built.

To get the project underway in 1922, Clarence Lewis and Helen Salomon chose John Russell Pope (1874-1937), then considered *the* preeminent Tudor architect, to design their stone and half-timbered residence.

The builder was the Elliot C. Brown Co. of New York City, which also built the country houses of Franklin D. Roosevelt in Hyde Park, N.Y., and E. Roland Harriman in Arden, N.Y.

Skylands Manor has 44 rooms on three floors, and a full basement. Astride a plateau on a northwest-southeast axis, it is surrounded on the north and east by the Ramapo Mountains, and open to the south and west. Though it gives every appearance of a 400-year-old English Tudor manor house, the imposing structure nevertheless has a solid modern core of steel and concrete with the very best heating, plumbing and engineering that money could buy in the 1920s.

An old view of the rear of Skylands Manor with the Octagonal Garden under construction in the foreground.

John Russell Pope

John Russell Pope (1874-1937) studied architecture at Columbia University, and went on to further his education at the American Academy in Rome and the École des Beaux Arts in Paris.

His first professional commission came in 1895, when he was chosen to design the memorial Chapel of St. Luke the Beloved Physician at Shepherd Lake.

Pope's public buildings include such well-known structures as the Jefferson Memorial and the National Gallery in Washington, D.C., and the Theodore Roosevelt Memorial at the American Museum of Natural History in New York. He was also selected to design additions to the Tate Gallery and the British Museum in London—a most unusual honor for an American architect.

Clarence Lewis and Helen Salomon selected Pope to design their Skylands Manor because his Tudor houses were probably the finest of the period, considered to be the most "correct" seen in this country up to that time. Among other Tudor manors Pope designed were those of Stuart Duncan in Newport, R.I.; Allen G. Lehman in Tarrytown, N.Y.; and William K. Vanderbilt in Great Neck, Long Island. But with all these accomplishments, Pope often cited Skylands as an example of his "very best residential work."

Pope believed in the reposeful serenity of the monumental classic, a belief reflected in his architectural style, and certainly displayed here.

Chapel of St. Luke at Shepherd Lake

The Manor is not a facsimile of any particular English estate, but a unique creation by Pope in the style of the period from late Gothic to the Renaissance, mostly 16th century. Its exterior includes such typical Tudor features as:

- large groups of rectangular windows,
- oriel windows (bay windows that do not extend to the ground floor level),
- intricate chimney complexes,
- crenelated walls,
- pargeting (decorative plaster work), and
- half-timbering.

The interior boasts such characteristic components as

- rich wood paneling,
- a Great Hall,
- molded plaster ceilings, and
- elaborately carved staircases.

Pope brought all these elements together during the Tudor Revival beginning early in the 20th century, producing impressive country houses for the wealthy. His classical training and inclinations gave the Manor House an elegant but austere, almost severe, appearance.

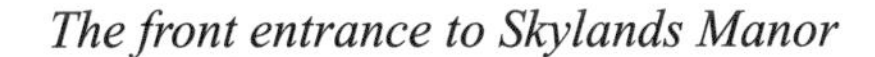

The front entrance to Skylands Manor

Skylands Manor

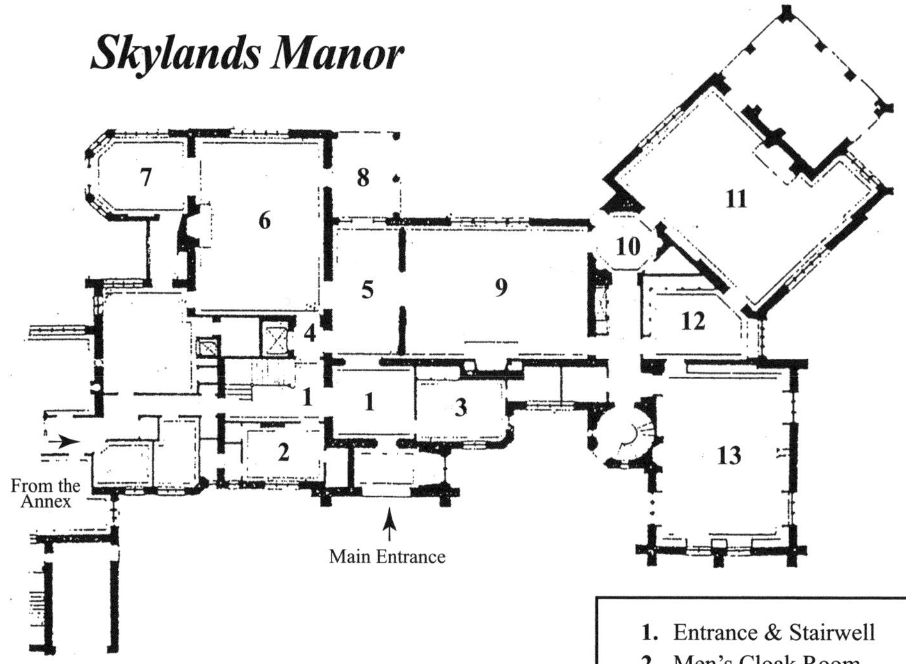

1. Entrance & Stairwell
2. Men's Cloak Room
3. Ladies' Cloak Room
4. Elevator
5. Center Hallway
6. Dining Room
7. Breakfast Room
8. Back Porch
9. Great Hall
10. Octagonal Hall
11. Drawing Room
12. Study
13. Library

The building is oblong with various end salients (projections), including a service wing with gabled roofs of different heights, a gabled library wing and a gabled west porch. Many gables have intersecting dormers.

The east face of Skylands Manor

The west face of Skylands Manor features half-timbering and lead downspouts

The foundation, first floor, main front, library wing and Great Hall exterior are native granite quarried at the Pierson Ridge area of Lewis's property, just above Emerald Pool. Here a supply with rectangular fractures was cut so that weathered surfaces could be used as the Manor House exterior, utilizing hundreds of years of natural weathering and coloring. Pope's blueprints specified to James McLaren & Sons, Cut Stone Contractors of Brooklyn, that "All stones must be accurately cut, marked section and course number showing the exact place where each belongs. The Stone Setter will be held responsible if stones are taken from where they belong to be put in any other place."

The remaining exterior walls are stucco (wattle and daub) and native oak timbers. The gable ends and other wood trim are oak, mostly carved by local artisans with elaborate patterns of grape vines, grouse and other flora and fauna.

The slate roof was specially designed with "waves" to simulate Tudor construction: first installing a wooden roof, then water-proofing,

Samuel Yellin

Samuel Yellin (1885-1940), who called himself "the blacksmith," led the 20th-century American revival of the use of iron as a decorative art.

Born in eastern Europe, he apprenticed there and had already achieved his master certificate when he came to America in 1906. By 1909 he had established his blacksmith shop in Philadelphia.

Yellin's first major commission, in 1911, was to produce an entrance gate for the Long Island estate of J. P. Morgan. At Skylands, he fashioned the lanterns, electrical fixtures, lamps, and spiral-staircase rail. But his designs and finished work here represent only a small portion of his accomplishments.

Along with his work at Skylands, he also designed the ironwork for the Princeton University Chapel, The Cloisters, St. Patrick's Cathedral, the Cathedral of St. John the Divine, and many other universities, institutions, churches, banks, and residential structures. One of his most impressive contracts was with the Federal Reserve Bank of New York, where he fashioned 200 tons of wrought iron, perhaps the largest such commission ever produced in the United States.

then laying plastered shingles at intervals, and finally covering it all with heavy slate. The slates were graduated, larger at first and rising to smaller at the roof tops, to achieve the appearance of a higher building elevation.

There are eight clusters of chimneys. The terra cotta chimney pots were coated with a mixture of cement and black paint to simulate age.

The individually-designed leaders and scuppers (leader-heads) appear to be all-lead antique Tudor fixtures, but they are actually copper covered with molten lead in an "unusual process," which Mr. Lewis said saved him $20,000.

All the exterior ironwork is made of non-rusting monel metal. The lanterns and electrical fixtures were designed by Samuel Yellin of New York and Philadelphia, considered among finest smiths ever to have worked in the United States.

Additional exterior decoration includes carved stucco panels, stone carvings over arches and doorways, carved corbels, and Gothic heads.

The entrance is at ground level, marked by the arched stone entry with the initials "C MK L" intricately carved in the door head. Above the entranceway is an oriel window on an elaborate stone corbel inset with the carving of a swan. The date 1924 also appears here, indicating that this section of the Manor was completed early on. Other sections of the exterior are also dated, allowing future archaeologists to establish an accurate chronology of construction.

The oriel window above the front entrance features a carved swan.

Similar corbeled oriel windows appear on the second story of the library wing, both in front and on its northwest side. Above the front oriel is a stone panel carved with an eagle and snake.

To the left of the main entrance is a central square crenelated tower enclosing the grand staircase. To the right is a circular tower containing a spiral staircase which was used by family and staff to easily access all floors. Gothic heads and foliage motifs of carved stone are placed at various intervals around these battlements.

Further to the south, to the left of the main entrance, the property drops down to a service court adjoining the half-timbered service wing. The rear façade, from the service wing around to the west porch, is half-timbered and includes stucco quatrefoils and various window groupings. The one exception here is the Great Hall exterior, whose seven arched windows and their 28 lights (panes) are entirely surrounded by stone ashlar.

Just below the leaderhead connected to the service wing at the building's south end we find the date 1925 and the architect's insignia, "JRP." Mrs. Salomon had intended a chapel to be included here, but during one of her absences Clarence Lewis re-purposed the location: To accommodate his beloved Pierce Arrow, he replaced the planned chapel with the gabled carport which juts out here. When she discovered this change on her last visit to Skylands, Mrs. Salomon was so distressed that she returned to New York City without getting out of her chauffeur-driven car.

Initially, she had consulted closely with Pope on every detail of the Manor's design. But over the years of construction she was kept away from Skylands more and more: first by buying trips, and then by declining health. Sadly, she died in 1927 without ever seeing the Manor fully completed.

Left: Helen Salomon
Right: The front entrance to Skylands Manor. The carport is on the left behind the tree.

The Lewis Dining Room's original furnishings provided table seating for as many as 12. This photograph, and others in this chapter, are from a series taken in 1929 by renowned architectural photographer Samuel H. Gottscho.

THE INTERIOR

As we prepare to enter, we see carved stone heads on each side of the entrance. The great oak door is decorated with owl heads and the initials "S" and "L" for Skylands.

The massive main entrance door opens to a stone-floored entrance foyer, and the connecting corridors at this level are also comprised of stone. But elsewhere in the structure the floors are mostly wood. Many floors throughout the house are pegged oak, while some parquet floors and borders include other types of wood.

The Gothic stone arches on the first floor display a variety of patterns, and little bats are carved into the stone of the archway leading to an Otis home elevator that runs the height of the building—one of the first such devices to be installed in a home anywhere.

The thistle, a symbolic reference to Clarence Lewis's Scottish heritage, is cut into the oak paneling above both sides of the archway facing the main entrance. Many other architectural touches both inside and outside the house reflect his wide-ranging interests.

The ceiling of the entrance foyer is reverse stamp-molded plaster that has been stained or painted to give the appearance of carved oak. The small window of this foyer contains one of Mr. Heinegke's thirteen unique glass pieces, and others can be seen in various places, mainly throughout the first floor of the house.

The entrance-hall woodwork is American oak, with carved decorative panels and friezes. The ceiling fixture and wall sconces are Mr. Yellin's work.

From the base of the main stairway, we can glimpse eleven pieces of the stained glass medallion collection in the arched windows facing the main landing. This stairway contains two flights leading to the second floor, and the balusters, newel post and caps have been intricately carved of American oak.

On one side of the main entrance is a ladies' hospitality room, displaying very feminine décor with a French flair, and a lavatory at its far end. In Lewis's time it offered guests a privacy screen in front of the lavatory door and a large, mirrored wardrobe closet.

The gentlemen's hospitality room is located under the large landing separating the flights of the main stairway. The room décor here is Tudor, simple and masculine, and provides two small closets and a lavatory.

Two doors open from the first floor center hall into the dining room. Its paneling and wall sconces were original to the Tudor House at Lyme Regis, Dorset, England. Mrs. Salomon bought these items from Charles Roberson, Knightsbridge Halls in London. The wood paneling of the room includes nineteen pilasters of exceptional quality, and the carved mantel is rich in de-

View from the entrance, facing the main stairway

Dining Room fireplace and plaster ceiling

tail. On either side of the overmantel are arcaded marquetry panels in the Elizabethan manner.

Pope chose to provide a Tudor ceiling of molded plaster here. Five pieces of the stained-glass medallion collection were placed in the window of this room. The window seat conceals the heating elements. (Throughout the house, most radiators are hidden from view, either under a window seat or behind a grille, to maintain the appearance of a manor house as it would have been in the sixteenth century.)

A butler's pantry, located behind the fireplace wall of the dining room and leading to the kitchen, serviced both the dining room and the adjoining breakfast room.

Now in the Dining Room, this table originally was in the Great Hall.

On the landing of the grand staircase at the front of Skylands Manor

Detail: St. George and the Dragon, one of the medallions in the Great Hall

Stained Glass

Large windows in each of the major rooms hold antique stained glass pieces dating back as far as 1554. Mostly of Swiss and German origin, they sparkle in sunshine and beckon visitors with a warm glow at night.

Above:
The Great Hall

Right:
The Dining Room

Below:
Upstairs Bedroom

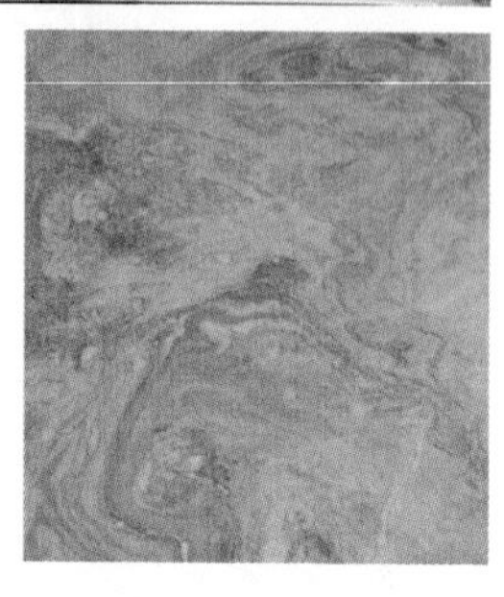

The Breakfast Room features walls of green marble and an exquisite Renaissance lavabo. Breakfast was served here promptly at 8:00 a.m., and lunch was served here, too, provided there were no more than six to be seated at table.

The walls of the breakfast room are lined with fine green Italian marble, while the floor is white marble. The windows afford a view of the terraced garden and the octagonal fountain. The dominant feature of the room is an ornate marble lavabo (ceremonial washbasin) from a Venetian palazzo. This basin came to Skylands from the Salomon mansion, and was kept filled with fresh flowers when the family was in residence. The molded plaster ceiling reflects the design of the lavabo. The room is lighted by a monastery censer that has been converted to electricity.

There are two porches on the first floor of the manor, both facing the West Terrace, a large walled section of lawn. From this area we can glimpse the distant Wanaque Reservoir, constructed during the first years of the Lewis occupancy.

The smaller of the two porches provides the connection between the dining room's west end and the central hallway. A large ceremonial bell hangs here, attesting to Mr. Lewis's interest in Asian art. Woodland animals grace the doorway stonework.

The Great Hall rises to the full height of this section of the building, and its nonstructural "jesting beams" of oak rest on carved stone corbels that resemble pelicans. This room is also constructed of American oak, with intricate carved panels. It is interesting to note the carved squirrels over the doorway and the acorn motif in the upper part of the panels.

The room is lighted by wall sconces as well as Samuel Yellin's chandeliers of wrought iron, in keeping with the period of the manor.

The windows of the Hall contain twenty pieces of the stained glass medallion collection and include depictions of Jonah and the Whale, St. George and the Dragon, and the Crucifixion, as well as the insignia of the cities of Berne, Switzerland, and Nuremberg, Germany.

The impressive mantelpiece is an exact replica of one from Over Manor, an estate in Gloucestershire. Including the royal coat of arms and dated 1619 (the period of James I),

In Lewis's day, the Great Hall had a tapestry over the fireplace. Notice the table under the window. One of just a few remaining original furnishings, it is in the Dining Room today.

it was duplicated by White Allom & Co., at whose London showroom Mrs. Salomon purchased it in 1925.

In Lewis's day, tapestries hung on the stone portions of the walls above the wood paneling.

An organ loft specially crafted for Mrs. Salomon is located above the massive oak doors leading from the main hallway. Sometimes referred to by the Lewis family as the "minstrels' gallery," it features very detailed carving reflecting a musical theme.

On the opposite wall, two doorways decorated with dragon heads lead into a connecting hallway. It extends from the round tower entrance at the manor's front, westward to the small Octagonal Hall, a space featuring display alcoves and an attractive overhead lighting fixture. This connecting hallway permits access to all levels of the manor via the circular stairway in the round front tower.

The door on the north side of the Octagonal Hall leads into the drawing room, which is faced with Scandinavian fir salvaged from Oulton Hall in Derbyshire, England. The carved mantelpiece trailing down to the marble fireplace is pearwood. Carved pediments top three of the drawing-room doors, and a carved "key of life" pattern surrounds the bookshelves.

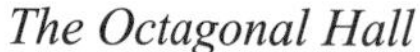

The Octagonal Hall

The Drawing Room

From this room one can exit to the larger of the two porches, which has a stone fireplace on its inside wall. This porch was formerly furnished with weathered teakwood benches and tables, built from an old ship.

The drawing room's west-facing window holds four more stained glass medallions. The small alcove at the eastern end of the room once contained a writing desk. Opposite this alcove is a door leading to the study.

The study, unlike other first-floor rooms, has a parquet floor. One of the smallest rooms of the manor, it is the most elaborately carved. The intricate panels were taken from a German Renaissance wardrobe and had been installed in Mr. Salomon's study in his New York City mansion. The priceless carvings found a third home at Skylands when the Fifth Avenue mansion was slated for demolition.

Mr. Lewis intended to use the study when working on his plant records. But, finding it to be too dark and small, he preferred to spend time in his library. The study's airtight cabinets were mainly used for storage.

In the 1990s this room underwent extensive restoration by a group of dedicated volunteers, under the direction of a master restorer. They stripped layers of varnish and dirt from the panels, revealing their lovely original wood tones and the exquisite quality of the carvings.

The Study

Most intricately carved of all the rooms in the Manor is the Study, which began life as a Renaissance wardrobe. Featuring birds, animals, horned dragon heads and other mythical monsters, its dark oak carvings were meticulously restored by a team of NJBG volunteers under the direction of a master restorer.

The library is entirely paneled in American oak, with a beamed ceiling and two of Samuel Yellin's chandeliers resembling those in the Great Hall. Here we see extensive bookshelves and cupboards, designed to accommodate Lewis's voluminous library of horticultural and general-interest books. The room's stern lines are somewhat tempered by the carved mantelpiece's pilasters, lozenges and inset arches.

Bordering the mantelpiece are glass shelves, which held a set of exquisite Japanese tea bowls; these were a "housewarming gift" to Lewis from his mother-in-law, an expert in Oriental art. In the glass-enclosed cabinets at the room's east end, Lewis displayed his prized collection of jade and other Oriental objects.

Regrettably, these art treasures, the Great Hall tapestries, and most of Lewis's fine furnishings are gone now. But many of the installed works by artisans from Heinegke to Yellin, from Venetian stonemasons to talented local woodcarvers, are still here as reminders of the Manor's grandeur in its glory days.

Grown at Skylands

Both Francis Stetson and Clarence Lewis were active in the botanical community in their day, and a number of plants are credited to them.

Above: Named for Clarence Lewis, Tsuga canadensis 'Lewis' *(behind rock) is a small, slow-growing evergreen. Its graceful and open branching habit gives it an airy look that contrasts nicely when planted near denser evergreens, such as here, near the bench in the Moraine Garden.*

*Right: The Skylands spruce is a cultivar of Oriental spruce (*Picea orientalis*). Small and slow-growing, it has both green and golden needles all year long and is particularly stunning nestled in winter snow. The tree shown at right is the original specimen from which all plants now in existence were grown.*

Left: Lewis introduced Hydrangea petiolaris 'Skylands Giant' *in 1952 through his affiliation with the New York Botanical Garden. Its white sterile flowers are much larger than most other hydrangeas. A lovely specimen blooms in front of Skylands Manor.*

Left: The Stetsonia coryne *is named after Francis Stetson. A native of Argentina, Bolivia and Paraguay, the 'Toothpick Cactus 'is the sole cultivar in this genus. This specimen is at the Huntington Library and Gardens in San Marino, California.*

The Botanical Garden

Clarence Lewis engaged the most prominent landscape architects of his day to design the gardens around his new summer home, and the results are still being enjoyed by thousands of people every year.

Most of the trees now framing the house were planted under Lewis's supervision, including the magnificent copper beeches. Lewis stressed symmetry, color, texture, form and fragrance in his gardens. He wanted to appeal to the senses. For thirty years, Lewis collected plants from all over the world and from New Jersey roadsides. The result is one of the finest collections of plants in the state.

Two majestic granite eagles grace the entrance to the Botanical Garden. They were brought here in the late 1960s, salvaged from New York City's Pennsylvania Station. Sculpted in 1906 by Adolph Alexander Weiman, each measures 56 by 64 inches.

Left: French in style, these statues in the corners of the Annual Garden are made of lead. Each statue contains elements of its season: winter, shown here, has a brazier and head covering.

Right: The statues on either side of the semi-circular bench, Greek in style, are reproductions. These waterbearers are of the Lewis period.

The Annual Garden Ovals are focused on two Fauns, reproductions of two lead statues that originally graced the alcoves of the Octagonal Hallway in the Manor House.

The Annual Garden

The layout of this Garden, like that in all the formal gardens at Skylands, has not been changed from the original design. It is the only garden at Skylands in which the main plantings are annuals. The displays in this garden therefore change not only through the seasons, but also from year to year. Note the small Four Seasons statues in the corners, and the Fauns which anchor the Ovals. Some of the benches here and in other parts of the garden were given by the NJBG/Skylands Association.

The centerpiece of the Annual Garden is the wellhead. This 16th century piece is also part of the East Vista toward Diana as seen from the main entrance of the Manor House. It is adorned with two lion heads and two cherub heads, joined by della Robbia garlands. Ropes used to draw water hundreds of years ago left grooves on the rim that are still visible today.

The All-America Garden

NJBG joins gardens all around the country as an AAS Display Garden, in a program sponsored by All-America Selections. Each year, the Garden's volunteers receive new seeds that are planted in this special bed, and the results are assessed nationally to determine which plants will receive the coveted 'All-America' designation.

Diana

Commonly known as 'Diana, Goddess of the Hunt,' or 'Diana de Versailles,' this is a copy in lead of a 4th century sculpture; the original is in the Louvre. Note the stag alongside Diana. This statue was cast around 1920 but is original to the garden. It was restored in 1993 by Paul Crowther, grandson of Henry Crowther, who originally cast the statue in Surrey, England. Funds for the restoration were donated by the NJBG/Skylands Association.

The Swan Boy. The original of this lovely statue was acquired by Helen Lewis Salomon on her European travels. This copy was imported from England.

The Perennial Border

Mr. Lewis's original cutting garden has been completely restored by NJBG volunteers. Lace vine covers the trellis surrounding the steps leading to this garden where a colorful floral display changes with the seasons. Intensive cultivation is required to maintain this garden.

There's always something on show in the Perennial Border, including (left to right) irises in the spring, goatsbeard in summer and beautyberry in the fall.

Clockwise from top left:

The Perennial Arbor is draped in lace vine.

Crocosmia

Buddha guards the passage between the Perennial and Annual Gardens.

Anemone

A superb place to soak up the summer sunshine and the colorful view.

Crabapple Vista

We have come to one of the Garden's most spectacular attractions, the Crabapple Vista. In early May, a profusion of pink blossoms stretches a half mile up the gentle slope to the Lodge.

At the south end of the Vista is the horse-chestnut collection, and to the east are Swan Pond and the Meadow, a field for moisture-loving plants such as flag iris, as well as some of Skylands's many varieties of willow. Higher parts of the meadow contain nut trees. The Vista marks the boundary between formal gardens to the west, near the Manor house, and the informal and Wildflower gardens to the east, at the foot of Mount Defiance here in the Ramapo Mountains.

Along the eastern edge of the meadow, note the planet signs for NJBG's scale-model Solar System, which stretches the length of the Vista. On this scale, the Earth is the size of a peppercorn.

In the Meadows

The meadows edging the Great Lawn are awash with irises in early summer, and bright yellow rudbeckia later in the season. The signposts of NJBG's scale-model solar system hug the eastern edge of the Lawn, and extend from the Lodge to the Four Continents.

Hosta & Rhododendron Garden

The fenced-in Rhododendron Garden also contains a large collection of hostas. These range in size from miniature to giant, and sport various shades of blue, green, and yellow-gold foliage, along with white and gold variegations. Surrounding them is a splendid collection of mature rhododendrons and azaleas. This garden is at its best in late May and June, and is fenced because these plants are a favorite snack food for the local deer population.

NJBG's hosta collection features more than three hundred varieties, including (clockwise from top left) Sieboldiana Mira, Paul's Glory, Sum and Substance, Patriot, and Grand Tiara.

The view of the Moraine Garden above shows it in its early days, and is much the same view as is shown below. Note the three tall evergreens, which still stand watch over this garden today, and have grown to massive specimens.

The Moraine Garden

New Jersey is home to many moraines, deposits of rock left behind by melting glaciers at the end of the last Ice Age. Mr. Lewis created this garden of mostly ground-hugging alpines, which thrive on rocky slopes with water seeping beneath. Look for heather, sedums, gentians, dwarf conifers, and many low spreading plants.

Japanese primroses line the paths of the Wildflower Garden in late May, offering a bold and jubilant welcome to spring.

The steppingstones on Bog Pond are a perennial favorite with young and old alike.

The Wildflower Garden

Winding wooded trails, steppingstone bridges and a frog-friendly Bog Pond make this part of the garden a favorite for youngsters. Native flowers and ferns are found throughout, with a beautiful display of Japanese primrose in late spring.

Many wildflowers carry fanciful names, such as (clockwise from top left) trillium, wild geranium, cardinal flower, spring beauty, gay wings and trout lily.

The Four Continents

These life-size limestone statues represent four of the continents—from left to right: Asia, America, Africa and Europe. The headgear and costume of each statue are easily identifiable. The grouping terminates the southern end of the Crabapple Vista.

Below: Archival photographs of the statues were part of the estate's original inventory, and include descriptive text in French on the back of each photograph. From left to right: Asia, America, Africa, and Europe.

The Woodlands

More than twenty-eight miles of roads and trails make exploring the hidden corners of the greater Botanical Garden a continual delight in any season.

Blooming from late April to early June, the Botanical Garden's Lilac Garden collection offers an excellent cross-section of the Syringa genus. It includes all seven major color groups and many different bloom types, as well as early and late bloomers. The profusion of blossoms and the delicious aroma draw thousands of visitors to the Garden each year.

The Lilac Garden

On the East Lawn, immediately adjoining the terraces, is the Botanical Garden's extensive lilac collection. It contains over one hundred varieties.

This garden is at its best near the middle of May, although some species will continue their bloom into June. The lilac's genus name, *Syringa*, is derived from the Greek word *syrinx* for "pipe," a reference to the hollow shoots. Lilacs belong to the olive family (*Oleaceae*) and therefore are related to white ash and privet. They are native to Europe and temperate Asia, where they grow as large shrubs or small trees.

Presumably, some lilacs predate Lewis at Skylands. Lilacs have been popular shrubs since Colonial times because of their ease of culture and their fragrant spring flowers. One of the first varieties to be recorded in Mr. Lewis's plant accession books is *Syringa x persica*, which he procured in 1923. In 1928, the Japanese tree lilac (*Syringa reticulata*), and the Chinese lilac (*Syringa x chinensis*) were purchased along with the French hybrids 'Edouard André' and 'Mme. Abel Chatenay.'

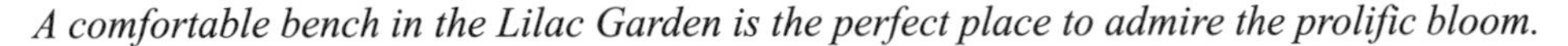

A comfortable bench in the Lilac Garden is the perfect place to admire the prolific bloom.

The Peony Garden

The Memory Bench at the far end of the Peony Garden is encircled by Canadian hemlocks (*Tsuga canadensis*). Family ashes were to have been placed in small vaults on each side of the bench, but they never were. Vandals have stolen the bronze plates that covered the vaults.

The tree peonies here are native to western China (called King of Flowers there). Unlike commonly known peonies, they are shrubby, with woody stems. Background plantings of deciduous flowering shrubs include weigela, mock orange, kolkwitzia and deutzia, which were popular in Victorian times.

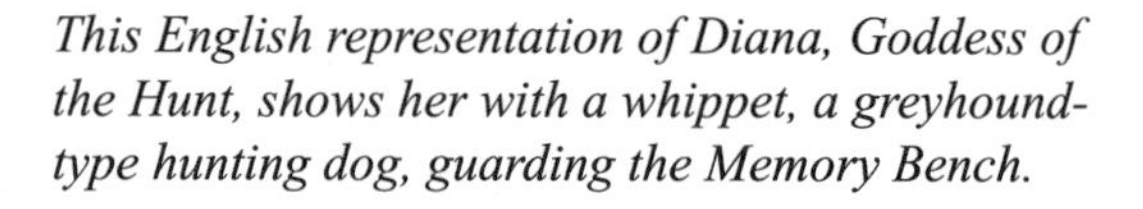

This English representation of Diana, Goddess of the Hunt, shows her with a whippet, a greyhound-type hunting dog, guarding the Memory Bench.

The Summer Garden

This lovely little area was originally the site of a rose garden. Air stagnation caused by the yew hedges made maintenance difficult, and the roses were replaced by daylilies. Because they are disease-resistant, daylilies need less care; they put on a colorful show during the summer months. Other summer annuals planted here vary from year to year, and you can count on a cheerful daffodil display every spring.

The Azalea Garden

Banks of azaleas and rhododendrons on both sides of the reflecting pool bloom in every conceivable shade. Hybrids include the white 'Boule de Neige' (French for snowball), 'Pink Twins' and the vivid red 'Nova Zembla.' The Japanese maple (*Acer palmatum 'dissectum'*) at the head of the pool and the globose sourwood (*Oxydendron arboreum 'globosum'*) are particularly striking in their crimson autumn foliage. Next to the maple is a mature mountain silverbell tree (*Halesia monticola*). In spring, look for the double flowering dogwood (*Cornus florida 'pluribracteata'*). In the summer, water lilies and koi grace the pool.

Below: A Carolina silverbell tree and dogwoods add even more color to the brilliant splash of the azaleas' pinks and mauves.

As you descend the terraced steps into the Azalea Garden, you will hear the splashing water of the Boy on the Dolphin tucked into the reflecting pool grotto. Although the original statue has long been gone, Clarence Lewis's daughter, Barbara, admired the grotto in the 1930s (top photo) and donated a replacement in recent years, The threadleaf maple on the right of the grotto is still growing there today.

Wild Visitors

The many different habitats at the Botanical Garden appeal to guests of all species. Birds, butterflies, and bees flit from flower to flower. Frogs and geese love the pools and ponds, deer wander the meadows, and black bears call the woodlands home.

Magnolia Walk

The sweet bay magnolias in Magnolia Walk are unusual because of their number and size this far north (they are a Southern species). Mr. Lewis planted them close to the house so that their sweet fragrance would drift into the manor windows in June. Note the many unusual shrubs on each side, including scented viburnums, honeysuckles and fragrant mahonias.

East of the walk is an unusual columnar form of the sugar maple, the sentry maple (*Acer saccharum monumentale*). And to the west is a Kentucky coffee tree. Its bark and very thick branches are interesting additions to the winter landscape. Also here are the Japanese pagoda tree (*Sophora japonica*), which blooms in August, and the fall-blooming golden-rain tree (*Koelreuteria paniculata apiculata*).

The Octagonal Garden

The rock garden surrounding the Octagonal Pool was designed so that small plants would be waist high, thus easily shown and maintained. It has been carefully restored, and its many dwarf plants are nicely displayed. Among them are *Sedum gypsicolum*, a creeping evergreen from Spain, the yellow-flowered *Friogonum flavum* from the mountains of the western United States, and monkshood (*Aconitum anthora*) from the Pyrenees.

The original yews between the stairways have grown out of proportion, but two Alberta spruces, also original, are still fine specimens because of their slow growth. On the east side of the courtyard is the Chinese toon tree (*Toona sinensis*), which, because of its interesting bark, open crown, and large white blooms, is an excellent companion planting with shrubs.

Opposite page: This unique old English lead statue was strategically placed in the pool in the Octagonal Garden. It was purchased by Mrs. Salomon in 1923.

The North Lawn

Offering expansive views over the surrounding hillsides, the North Lawn is framed in cascades of cherry blossoms in late April.

The Winter Garden

Lewis planted much of this garden in 1927-1928, but the red oak in front of the manor house library window overlooking the Winter Garden dates back to the 1890s. At that time it was surrounded by Stetson's nine-hole golf course. Mr. Lewis's Winter Garden is a collection of forms, textures and colors to stimulate the senses in winter. Notice the golds, blues and reds among the many evergreens.

On the west side of the Winter Garden is New Jersey's largest Jeffrey pine (*Pinus jeffreyi*). The east side is dominated by a weeping beech next to an upright beech that is a century old, planted by Stetson. The Japanese umbrella pine (*Sciadopitys verticillata*) is one of the most distinctive and handsome conifers at Skylands, and was planted by Stetson sometime between 1891 and 1920.

Other interesting trees include an Algerian fir (*Abies numidica*), which was grown from seed in 1931. The Atlas cedar (*Cedrus atlantica*), also a native of North Africa, is distinguished by its graceful appearance, erect cones and clusters of one-inch-long needles borne on spur shoots. The Atlas cedar growing in the Winter Garden was purchased by Mr. Lewis in April 1928. The less-erect blue type can be found next to the green form.

The Winter Garden, with its wide range of evergreen colors, shapes and textures, is lovely in all seasons. The Jeffrey pine can be seen peeking out in the center background.

In early spring, the statue of Diana is framed by a lush display of cherry blossoms.

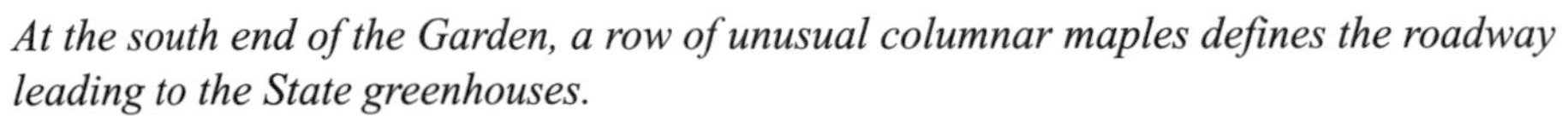

At the south end of the Garden, a row of unusual columnar maples defines the roadway leading to the State greenhouses.

Japanese primroses in the Wildflower Garden

Acknowledgements

This book owes its existence to many persons and organizations, beginning with the generous financial support provided by the Fred J. Brotherton Charitable Foundation and the members of the NJBG/Skylands Association.

We appreciate the help of representatives of the State of New Jersey, and in particular would like to thank Governor Thomas H. Kean, Senator Leanna Brown, and Congressman Robert Roe, whose foresight created the Botanical Garden we all enjoy today.

Many thanks are also due to the NJBG Book Committee and the enthusiastic friends of the Botanical Garden who contributed their knowledge, their time, and their talents to making this book a reality.

NJBG Book Committee work session (left to right): Thomas Grissom, Dorothy Dobek, Richard Flynn, Richard Cahayla-Wynne, Andrew Noll, and Alex Rainer. Maja Britton is behind the camera.

Photo Credits

Bigliano, Eugene – *57b* • Bristow, Nancy – *95tl, 95tr, 95cl, 95cr, 95bl, 95br, 111t* • Britton, Maja – *Dustcover front; 05, 10, 14t, 20b, 52t, 53, 54t, 55t, 56t, 74cr, 80r, 80cr, 80t, 81, 82b, 86t, 89c, 90, 92b, 93, 94b, 97t, 97b, 102b, 105, 106, 116* • Brown, Sen. Leanna – *44, 45b, 46b, 48b, 48t* • Carnegie Library of Pittsburgh – *26tr* • David & Lorraine Cheng Library, William Paterson University of New Jersey – *46t* • Cooper, Joseph – *14b, 63b, 65t, 65c, 66bl, 67b, 70b, 71b, 71tl, 71tr, 72b, 72t, 73b, 73c, 73t, 74tl, 75tr, 76b, 76t, 77bl, 77tr, 78br, 78cr, 78l, 78tr, 79b* • Freeman, Daiv, cactiguide.com – *80b* • Fuda, Melina – *09, 82cr, 87b, 96t, 100b,100t, 101b, 101t, 102c, 104tl, 104tr, 107b, 107t, 110b, 110t, 114* • Gall, John – *91tl, 91tr, 91cl, 91cr, 91bl, 91br* • Grissom, Thomas – *06, 38c, 47b, 49b, 54b, 59c, 59t, 100c, 109* • Hawthorne, Gail – *07, 87br, 87tl, 90t, 94t, 98bl, 102t, 104c* • Kaar, Edwin – *87tr, 104b, 106br, 113b* • Kean, Thomas H. – *8* • Langer, Ingeborg – *86bl, 86bc, 86br* • Library of Congress – *17, 26tc, 26tl, 69, 74tr, 75bl, 75br, 77tl, 79cr* • Maniscalki, Pauline – *11, 55b, 56b, 112t* • Merkel, Diana – *57c, 85* • New Jersey State Archives / Skylands Collection – *12t, 16, 19, 20t, 22b, 22t, 24b, 25b, 25t, 27b, 28tl, 28tr, 29tl, 30, 31, 32, 34, 36t, 37b, 37c, 38b, 38t, 39b, 39t, 40, 41, 42b, 42t, 43bc, 43bl, 43br, 43cr, 43tl, 61, 62t, 64b, 68bl, 92t, 96b, 103t* • NJBG – *18bl, 29tr, 33br, 33cr, 33tr, 35, 64t* • NJBG/Alfred Hopkins – *23b, 24t* • North Jersey Media Group / Joe Sarno – *51b*, • Piff, Maria – *29b, 36c, 50t, 50b, 52b, 57t, 58b, 59b, 60, 62b, 65b, 84, 87cr, 88, 89t, 89b, 108, 111b* • Platt, Peter – *83* • Rainer, Alex – *68br, 82t, 82cl, 103b, 112b* • Rounds, Sharon – *49t* • Wallace, Dr. Edith – *80c, 98br, 98cl, 98cr, 98t, 99* • Wyckoff, Jerome – *12b, 15, 58t* • Yellin, Clare (courtesy of Samuel Yellin Metalworkers Co.) – *66tl, 66cr*

Timeline

1891	Francis Lynde Stetson buys the 300-acre Storms (Brookside) Farm
1892	Stetson buys 200 acres of the 400-acre Morris Farm
1892	Helen Forbes Lewis marries William Salomon
1897	The Stetson manor house is completed
1905	Stetson buys "the Brush Place," 170 acres around Brushwood Pond
1907	Stetson buys his last parcel – 300 acres around Pierson Ridge – from the Ramapo Manufacturing Company
1908	Clarence Lewis marries Annah Churchill Ripley
1912	Clarence and Annah Lewis buy Sheffield Farm in Mahwah
1918	Annah Lewis dies
1919	William Salomon dies
1920	Francis Lynde Stetson dies on December 5th
1922	Skylands deed to Lewis filed April 16th
1922	Lewis chooses John Russell Pope to design his new manor
1923	Lewis buys the adjoining Hillair estate
1927	Helen Forbes Lewis Salomon dies
1928	The Lewis manor house is fully completed
1953	Lewis sells Skylands to the National Bible Institute, which uses it as the site of its Shelton College
1959	Clarence Lewis dies in New York City
1966	New Jersey buys Skylands as the first purchase under the Green Acres program
1976	NJBG/Skylands Association, Inc. is founded
1985	Gov. Thomas H. Kean dedicates Skylands as New Jersey's official botanical garden
1990	Skylands is listed on the New Jersey Register of Historic Places
1992	Skylands is listed on the National Register of Historic Places

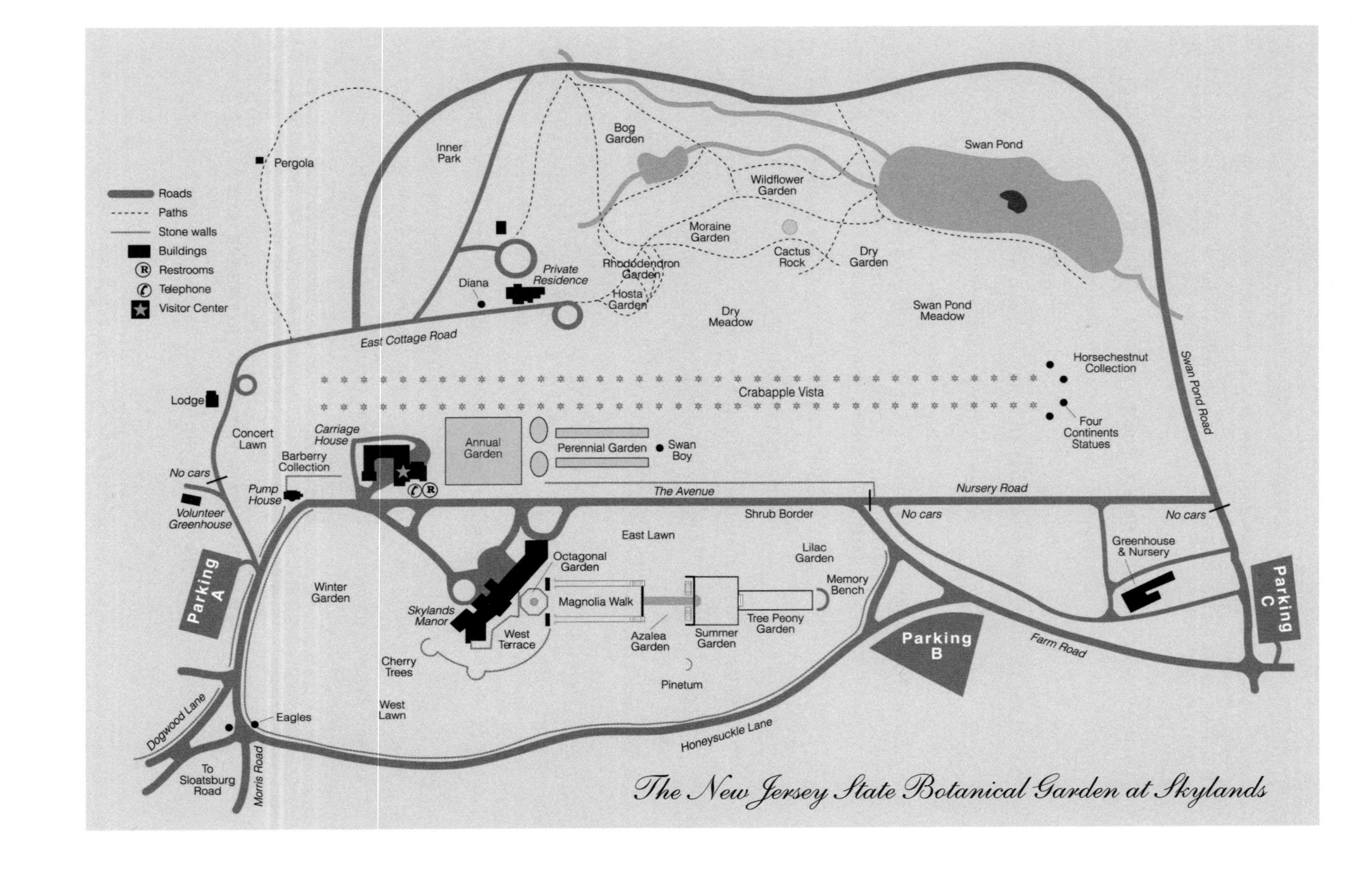

Roads
Paths
Stone walls
Buildings
Restrooms
Telephone
Visitor Center
Pergola
Inner Park
Bog Garden
Swan Pond
Wildflower Garden
Moraine Garden
Cactus Rock
Dry Garden
Private Residence
Rhododendron Garden
Diana
Hosta Garden
Dry Meadow
Swan Pond Meadow
East Cottage Road
Horsechestnut Collection
Swan Pond Road
Lodge
Crabapple Vista
Four Continents Statues
Concert Lawn
Carriage House
Annual Garden
Perennial Garden
Swan Boy
Barberry Collection
No cars
Pump House
The Avenue
Nursery Road
Volunteer Greenhouse
Shrub Border
No cars
No cars
East Lawn
Lilac Garden
Greenhouse & Nursery
Parking A
Octagonal Garden
Winter Garden
Memory Bench
Magnolia Walk
Skylands Manor
Tree Peony Garden
Parking C
West Terrace
Azalea Garden
Summer Garden
Parking B
Farm Road
Cherry Trees
Pinetum
West Lawn
Eagles
Dogwood Lane
Honeysuckle Lane
To Sloatsburg Road
Morris Road
The New Jersey State Botanical Garden at Skylands